THE PLANET NEPTUNE
An Historical Survey Before Voyager
Second edition

WILEY-PRAXIS SERIES IN ASTRONOMY AND ASTROPHYSICS
Series Editor: John Mason, B.Sc., Ph.D.

Few subjects have been at the centre of such important developments or seen such a wealth of new and exciting, if sometimes controversial, data as modern astronomy, astrophysics and cosmology. This series reflects the very rapid and significant progress being made in current research, as a consequence of new instrumentation and observing techniques, applied right across the electromagnetic spectrum, computer modelling and modern theoretical methods.

The crucial links between observation and theory are emphasised, putting into perspective the latest results from the new generations of astronomical detectors, telescopes and space-borne instruments. Complex topics are logically developed and fully explained and, where mathematics is used, the physical concepts behind the equations are clearly summarised.

These books are written principally for professional astronomers, astrophysicists, cosmologists, physicists and space scientists, together with post-graduate and undergraduate students in these fields. Certain books in the series will appeal to amateur astronomers, high-flying 'A'-level students, and non-scientists with a keen interest in astronomy and astrophysics.

ROBOTIC OBSERVATORIES
Michael F. Bode, Professor of Astrophysics and Assistant Provost for Research, Liverpool John Moores University, UK

THE AURORA: Sun–Earth Interactions
Neil Bone, School of Biological Sciences, University of Sussex, Brighton, UK

PLANETARY VOLCANISM: A Study of Volcanic Activity in the Solar System, Second edition
Peter Cattermole, formerly Lecturer in Geology, Department of Geology, Sheffield University, UK, now Principal Investigator with NASA's Planetary Geology and Geophysics Programme

DIVIDING THE CIRCLE: The Development of Critical Angular Measurement in Astronomy 1500–1850
Second edition
Allan Chapman, Wadham College, University of Oxford, UK

THE DUSTY UNIVERSE
Aneurin Evans, Department of Physics, University of Keele, UK

COMET HALLEY - Investigations, Results, Interpretations
Volume 1: Organization, Plasma, Gas
Volume 2: Dust, Nucleus, Evolution
Editor: John Mason, B.Sc., Ph.D.

ELECTRONIC AND COMPUTER-AIDED ASTRONOMY: From Eyes to Electronic Sensors
Ian. S. McLean, Department of Astronomy, University of California at Los Angeles, California, USA

URANUS: The Planet, Rings and Satellites
Ellis D. Miner, Cassini Project Science Manager, NASA Jet Propulsion Laboratory, Pasadena, California, USA

THE PLANET NEPTUNE: An Historical Survey Before Voyager, Second edition
Patrick Moore, CBE, D.Sc.(Hon.)

THE HIDDEN UNIVERSE
Roger J. Tayler, Astronomy Centre, University of Sussex, Brighton, UK

Forthcoming titles in the series are listed at the back of the book.

Table of Contents

List of Plates

Plate no.

Note: Position of Plate Sections
Plates 1 to 16 inclusive appear between pages 42 and 43
Plates 17 to 22 inclusive appear between pages 82 and 83 (colour section)
Plates 23 to 30 inclusive appear between pages 92 and 93

THE PLANET NEPTUNE

An Historical Survey Before Voyager
Second edition

Patrick Moore, C.B.E., D.Sc.

JOHN WILEY & SONS

Chichester • New York • Brisbane • Toronto • Singapore

PRAXIS

Published in association with
PRAXIS PUBLISHING
Chichester

Copyright © 1996 Praxis Publishing Ltd
The White House,
Eastergate, Chichester,
West Sussex, PO20 6UR, England

First published in 1988
This Second edition published in 1996 by
John Wiley & Sons Ltd
in association with Praxis Publishing Ltd

Wiley Editorial Offices

John Wiley & Sons Ltd, Baffins Lane,
Chichester, West Sussex PO19 1UD, England

John Wiley & Sons, Inc., 605 Third Avenue,
New York, NY 10158-0012, USA

Jacaranda Wiley Ltd, G.P.O. Box 859, Brisbane
Queensland 4001, Australia

John Wiley & Sons (Canada) Ltd, 22 Worcester Road,
Rexdale, Ontario M9W 1L1, Canada

John Wiley & Sons (SEA) Pte Ltd, 37 Jalan Pemimpin 05-04,
Block B, Union Industrial Building, Singapore 2057

A catalogue record for this book is available from the British Library

ISBN 0-471-96015-2

Printed and bound in Great Britain by Hartnolls Ltd, Bodmin

Preface to the first edition

In this book I have done my best to give a concise summary of our present knowledge of the outermost giant planet, Neptune. I hope that it will be of some interest to the general reader, and that the lists of references, at least, will be of some value to the more serious student.

My grateful thanks are due to Dr D. W. Dewhirst, of Cambridge University, for providing me with lists of the important papers in the University archives which relate to the discovery of Neptune; to Mrs Norma Foster, a descendant of John Couch Adams, for making available to me some unpublished manuscripts; to the Royal Astronomical Society, for allowing me to reproduce extensive quotes from their publications; to Paul Doherty, for his skilful illustrations; and to the publishers, Ellis Horwood Ltd., particularly to Felicity Horwood for her help and encouragement.

Selsey
October 1988

Patrick Moore

Preface to the second edition

When the first edition of this book appeared, shortly before the Voyager 2 encounter with Neptune, it was — so far as I knew — the only full account of what we had then found out about the outermost giant plant. I did my best to index all the published papers of any importance, and the fact that I could do so in less than a hundred and fifty pages shows how meagre our knowledge was at the time.

Since the Voyager mission, things have changed, and the published papers about Neptune now amount to many thousands. Moreover, many of our former ideas have been proved wrong; Neptune does have a ring system, there are no seas on Trition, and half a dozen new satellites have been found. Moreover, the detection of small bodies in the hypothetical "Kuiper Belt" has changed many of our views about the nature of Triton and even Pluto.

What I have done, therefore, is to leave the original text almost untouched, but at the end of most chapters I have added, in a new final paragraph, a note about how the discoveries of Voyager have modified our views. I have than included an extended Epilogue (Chapter 13) which gives a necessarily brief account of the Voyager findings. I must stress, therefore, that this book remains essentially historical, and does not set out to present anything in the nature of a full description of the post-Voyager Neptune.

I am most grateful to the staff of the libraries of the Royal Society, Royal Astronomical Society, and Institute of Astronomy, Cambridge, and to the Jet Propulsion Laboratory and Space Telescope Science Institute for assistance with the photographs, and to Clive Horwood and John Mason for their help with the preparation of this new, second editon.

Selsey
September 1995

Patrick Moore

1

Introduction: The planetary system

The planet whose position you have pointed out actually exists. The same day that I received your letter, I found a star of the eighth magnitude which was not shown on the excellent chart (drawn by Dr Bremiker), *Hora XXI* of the series of celestial maps published by the Royal Academy of Berlin. The observations made on the following day determined that this was the sought-for planet.

So wrote Johann Galle, astronomer at the Berlin Observatory, to the French mathematician Urbain Le Verrier on 25 September 1846. Le Verrier's calculations, and the careful observations of Galle and his young assistant Heinrich D'Arrest, had added a new planet to the Solar System: the world we now call Neptune. Yet surely none of those concerned in the discovery could have foreseen that in less than a century and a half a man-made probe, Voyager 2, would pass within a few thousand kilometres of the new planet.

Neptune is a long way away. Its mean distance from the Sun is 4497 million kilometres, and it has a revolution period of 164.8 years. Even though it is much larger than the Earth, it is too faint to be seen with the naked eye, and not even our largest telescopes will show any conspicuous markings upon its tiny, bluish disk. At the moment, at least, it marks the boundary of the known planetary system.

This is no place to give anything like a detailed description of the Sun's family; there are many books which do so (for example, Greeley (1987)). Therefore, I propose to confine myself almost entirely to matters which are of direct relevance to a discussion of Neptune.

The Solar System is made up of one star (the Sun), the nine known planets, the satellites of the planets, and various minor bodies such as asteroids, comets and meteoroids, together with a large amount of interplanetary gas and 'dust'. Any casual glance at a plan of the system will show that it is divided into two definite parts. First we have four relatively small, solid planets: Mercury, Venus, the Earth and Mars. Then comes a wide gap, in which move the thousands of asteroids or minor planets, of which only one (Ceres) is as much as 1000 kilometres in diameter and only one (Vesta) is ever visible without optical aid. Beyond lie the four giant planets Jupiter, Saturn, Uranus and Neptune, together with Pluto, a curious little body which does not seem to fit into the general pattern, and which may not be worthy of true planetary status.

The Earth's age is known, with reasonable accuracy, to be about 4.6 thousand million years. It is reasonable to assume that the other members of the Solar System are of roughly the same age — and we can prove this in the case of the Moon, because we have been able to analyze lunar samples in our laboratories. There seems no doubt that the planets were formed from a cloud of material associated with the embryo Sun, and that they grew, by accretion, from this 'solar nebula'. Volatiles were driven away from the planets which formed close-in; conditions in the outer regions of the system were different, and so the giant planets could retain their lighter gases, notably hydrogen. This can explain why the two parts of the Solar System are so different.

Since the start of the Space Age, with the launch of Russia's Sputnik 1 on 4 October 1957, we have sent unmanned probes out to the planets, and soft landings have been achieved on both Venus and Mars, though as yet only the Moon has been reached by man. It is fair to say that our knowledge of the Solar System has increased beyond all recognition over the past thirty years, and many of our long-cherished theories have to be abandoned; no doubt there are many more surprises in store for us during the coming decades.

The four inner planets have various points in common. All are rocky, and all are presumably made up of much the same materials. The differences between them with regard to surface conditions are due partly to their different sizes and masses, and partly to their different distances from the Sun. Thus Mercury, only 4878 km in diameter, has a low escape velocity, and has been unable to hold on to any appreciable atmosphere; as we know from the images sent back by Mariner 10, the only probe to have by-passed it so far, it is mountainous and cratered. Superficially, the Mercurian landscape looks very like that of the Moon, though there are differences in detail. It is dense, with a large iron-rich core, and there is a detectable magnetic field. It is a slow spinner: it takes 58.7 Earth days to turn once on its axis — a period to equal to two-thirds of a Mercurian year.

Venus, almost a twin of the Earth in size and mass, has a dense atmosphere made up chiefly of carbon dioxide. The surface temperature is very high (around 480°C), and the clouds contain large amounts of sulphuric acid, so that conditions there are unsuitable for any form of life as we know it. Both Russian and American automatic probes have mapped the surface by radar, and we even have images transmitted direct from the surface. There are two main highland areas, with lofty mountains; it is also thought that there are active volcanoes, though we cannot be sure. Unlike Mercury, Venus has no detectable magnetic field.

Mars is less unwelcoming. Though it is small, with a diameter only about half that of the Earth, it has retained an appreciable atmosphere, though the main constituent is again carbon dioxide and the ground pressure is everywhere below 10 millibars. The space-craft which have by-passed it show that it is mountainous and cratered, with towering volcanoes which are probably extinct. The Viking landers of 1976 have shown no signs of life, and we may be sure that if there are any living things on Mars they must be very lowly. The dark areas, once thought to be old sea-beds filled with organic material, have proved to be nothing more than albedo features.

Where the inner planets differ most strongly is in their satellite systems — or lack of them. Mercury and Venus are unattended. The Earth, of course, has one satellite — the Moon — which is probably worth classifying as a companion planet, since it has 1/81 of the mass of the Earth. The Moon has, to all intents and purposes, no atmosphere, since the escape velocity is a mere 2.38 kilometres per second, and analyses of the lunar rocks have shown that there is no hydrated material. The Moon has never had any water, and neither can it have had any life. The surface is dominated by the grey plains still mis-called 'seas', together with mountains, peaks, valleys, and of course the walled plains and craters, which range from tiny pits up to vast enclosures well over 250 km in diameter. Most astronomers (though not all) believe that the craters were produced by meteoritic bombardment rather than by internal action.

Mars has two dwarf satellites, Phobos and Deimos, neither of which is as much as 30 km in diameter. They are irregular in shape, with cratered surfaces; both have been surveyed by American space-craft, and have proved to be quite unlike our Moon. There is at least a chance that they are ex-asteroids which were captured by Mars in the early history of the Solar System.

The inner planets are relatively close to us, and so our present-day rockets can reach them in periods of a few months. However, the only probe to have been sent to Mercury (Mariner 10) used the so-called 'gravity-assist' technique. Launched in November 1973, it by-passed Venus in February 1974, and used the gravitational pull of that planet to send it on to a rendezvous with Mercury in the following month. This sort of procedure has been of immense importance in the exploration of the outer planets, particularly Uranus and Neptune.

The asteroids, most of which move between the orbits of Mars and Jupiter, were once thought to be the débris of a larger planet (or planets) which broke up, but this idea has fallen into disfavour, and it now seems more likely that no large planet could form in this region because of the powerful disruptive pull of Jupiter, which is much the most massive member of the Sun's family. Thousands of asteroids are known, and no doubt there are many more, too small and faint to be seen from Earth. When the first probes were sent to Jupiter and beyond, there were serious qualms about the dangers of passing through the asteroid zone; a collision between a space-craft and even a tiny asteroid would be disastrous. However, the six space-craft which have so far passed through (Pioneers 10 and 11, Voyagers 1 and 2, Galileo and Ulysses) have emerged unscathed, so that either the danger is less than was originally believed, or else we have been remarkably lucky.

Some asteroids swing away from the main zone, and may come close to the Earth; some (such as Icarus and Phæthon) even invade the regions closer to the Sun than the orbit of Mercury. All these are very small, but the Trojans, which move in the same orbit as Jupiter, are more substantial; at least two are almost 200 km in diameter. We also have the exceptional Chiron, which has been given an asteroidal number (2060) but moves mainly between the orbits of Saturn and Uranus. Its precise nature is still uncertain.

The giant planets are very 'spread-out'. A modern-type rocket takes years to reach them, and the gravity-assist technique has been invaluable. It so happened that during the late 1970s the four giants were arranged in a curve, which enabled Voyager 2 to visit each in turn — a situation which will not recur for well over a century. Voyager 2 was launched in 1977, and took less than 12 years to rendezvous with Neptune, having made successful passes of Jupiter (1979), Saturn (1981) and Uranus (1986). Without the gravity-assist procedure, the journey to Neptune would have taken much longer.

Jupiter and Saturn are essentially similar, though Jupiter is considerably the larger and more massive of the two. It is thought that Jupiter has a silicate core at a temperature of the order of 30 000°C; around this is a deep shell of liquid metallic hydrogen, which is in turn surrounded by a shell of liquid molecular hydrogen. Above this comes the gaseous atmosphere, which is of the order of 1000 km deep, and is made up of 89 per cent hydrogen, 10 per cent helium and 1 per cent of other elements. It contains water droplets, ice crystals, ammonia crystals and ammonium hydrosulphide clouds at various levels. When we look at Jupiter we are seeing not a solid surface, but merely a cloud layer; the temperature of the cloud tops is around −150°C. Telescopes show dark belts, bright zones, and other features of which the most famous is the Great Red Spot.

Jupiter radiates 1.7 times more energy than it would do if it depended entirely upon radiation received from the Sun, so that there is considerable internal heat. There is strong radio emission, and a very powerful magnetic field. There are also zones of radiation which would be quickly lethal to any astronaut foolish enough to venture inside them.

Most of our knowledge of Jupiter has been obtained from four probes which have passed by it, Pioneer 10 (1972), Pioneer 11 (1973) and the two Voyagers (1979). Both Voyagers used the gravity-assist technique to send them on to rendezvous with Saturn, in 1980 and 1981 respectively.

Saturn, like Jupiter, seems to have a silicate core surrounded by layers of liquid hydrogen, but the outer clouds contain more hydrogen and less helium than those of Jupiter. The core temperature is lower, and the mean density of the globe is less; it is actually no more than 0.7 that of water. Like Jupiter, Saturn has a marked internal heat-source. Both these giant worlds are in rapid rotation, and their globes are appreciably flattened, as any telescope will show.

Saturn's magnificent rings are made up of icy particles. The Voyagers have shown them to be immensely complex, with thousands of ringlets and narrow divisions. Jupiter, on the other hand, has only a thin, dark ring not detected until the Voyager 1 pass. The satellite families also are very different, and this is relevant when we come to consider the satellites of Neptune.

Of Jupiter's sixteen known satellites, twelve are small (below 300 km in diameter) and of these the outer four, which are true midgets, may be captured asteroids; their orbital motions are retrograde. However, the remaining four satellites, known as the Galileans because they were first studied systematically by the Italian astronomer Galileo in 1610, are of planetary size. Europa is slightly smaller than our Moon, Io slightly larger, and Ganymede and Callisto much larger. Indeed, Ganymede has a diameter greater than that of Mercury, though its mass and density are less.

All four Galileans were surveyed by the Voyagers, and have proved to be extraordinary worlds. Ganymede and Callisto are icy and cratered; Europa has a smooth, icy surface, while Io is sulphur-covered, with violently active sulphur volcanoes. Io also has a pronounced effect upon Jupiter's radio emission. However, none of the Galileans has an appreciable atmosphere.

Saturn's family is dominated by Titan, which is almost as large as Ganymede (diameter 5150 km) and has a dense atmosphere, consisting mainly of nitrogen, argon and methane. The ground pressure is 1.6 times that of the Earth's air at sea-level. Voyager 1 by-passed Titan at a distance of only 6490 km, but the images showed nothing more than the top of a layer of orange 'cloud', and we are still uncertain about the surface conditions. The globe may be about 55 per cent rock and the rest ice. The surface temperature is about −180°C, which is near the triple point of methane; it has been suggested that there may be cliffs of solid methane, rivers of liquid methane and a constant rain of organics from the orange clouds. Alternatively, there may be a methane ocean. Titan is the only satellite in the Solar System known to have an appreciable atmosphere.

The other satellites of Saturn are smaller. Rhea, Iapetus, Dione and Tethys have diameters between 1600 and 1000 km; all these have icy, cratered surfaces, though Iapetus is unusual in having hemispheres of unequal reflecting power. Of the rest, Mimas and Enceladus have diameters of around 400 km and 500 km respectively; Hyperion is irregular in form (its shape has been likened to that of a hamburger) and the rest are very small indeed. Phœbe, the outermost satellite, has retrograde motion, and, like the outer members of Jupiter's family, may well be asteroidal.

Voyager 1 was programmed to survey Titan as well as Saturn itself, and this meant that it was unable to move on to rendezvous with Uranus and Neptune. This was left to Voyager 2, which reached Uranus in January 1986 before travelling on to an encounter with Neptune in August 1989.

Uranus and Neptune are almost twins; Uranus is slightly the larger, but is the less massive of the two. But though Uranus and Neptune make up a pair, they are very different from the Jupiter–Saturn pair. They are much smaller, though they still rank as giants, and they contain much more of substances such as water and ammonia. Moreover, they are in some ways unlike each other. Uranus has an exceptional axial inclination of 98 degrees, more than a right angle, so that there are times when one or other of the poles is turned toward the Sun and the Earth. The Uranian calendar is very strange: during each Uranian 'year', 84 times as long as ours, each pole has a 'night' lasting for 21 Earth years and a 'day' of equal length. The reason for this state of affairs is not known. It has been suggested that Uranus suffered a major collision with another body early in its history, but the whole question is still very open.

There are other differences, too. Neptune has a source of internal heat; Uranus has not. Uranus has a system of thin, dark rings, and it has five moderately large satellites (Miranda, Ariel, Umbriel, Titania and Oberon) while Neptune has only one (Triton). The Uranian satellites are icy. Craters abound on Ariel, Titania and Oberon, and there are craters also on the darker and more subdued surface of Umbriel, while Voyager 2 showed that Miranda has an incredibly varied landscape with craters, cliffs, and large areas which have been nicknamed 'race-tracks'. Uranus also has ten small inner satellites, detected from the equipment carried on Voyager 2.

Triton, Neptune's only large satellite, has retrograde motion, while all the other known retrograde satellites in the Solar System are small and probably asteroidal. Nereid, the other satellite of Neptune known before the Voyager 2 pass, is small, and has a highly eccentric orbit, more like that of a comet than a satellite. It has direct motion, but the whole situation is most peculiar, and has led to the suggestion that something cataclysmic happened in the area long ago — perhaps even involving the remaining known planet, Pluto.

Pluto is itself an oddity. It is smaller than our Moon, with a diameter of about 2325 km, and it is accompanied by a similar body, Charon, with about half Pluto's diameter. The revolution period of Charon (6.387 Earth days) is the same as that of Pluto's axial rotation, so that to an observer on Pluto, Charon would appear fixed in the sky. Moreover, Pluto's path brings it within that of Neptune, though the orbital inclination of 17.14 degrees means that there is no danger of collision. The revolution period is nearly 248 years. The date of the current perihelion is 1989, so that between 1979 and 1999 Pluto is closer-in than Neptune.

Pluto's small size, plus the presence of Charon and its exceptional orbit, may show that it is not a true planet. It has even been suggested that it was once a satellite of Neptune which broke free and moved away independently; if so, then the same sequence of events may have thrown Triton into a retrograde orbit and Nereid into a highly eccentric one. There are serious drawbacks to this idea, but the possibility of a former association between Neptune and the Pluto–Charon pair remains.

Finally we must take note of the comets, which are of low mass, and most of which have very elongated orbits. There are many periodical comets with periods of a few years or a few tens of years, but only one of these can ever become bright: Halley's, which has an average return period of 76 years and which last passed through perihelion in 1986. At aphelion, Halley's Comet moves out well beyond the orbit of Neptune, and efforts have been made to see whether it is being perturbed by a still more distant planet which remains to be discovered — a point to which I will return later. All other periodical comets are relatively faint. As they lose material by evaporation from the icy nucleus every time they pass through perihelion, they are short-lived by cosmical standards, and several comets which were seen regularly during the last century have now disappeared. One of these, Biela's Comet, played a rôle during the search for Neptune in 1846. The brilliant comets seen at unpredictable intervals have periods so long that we cannot calculate them with any accuracy; it is thought that they come from a 'cloud' of such objects moving at a distance of at least one light-year from the Sun. We cannot prove the reality of this Oort Cloud, named in honour of the Dutch astronomer Jan Oort, but it does seem to fit the facts.

Such is our Solar System. Of the planets, five are naked-eye objects, and have been known since the dawn of human history. Of the rest, Uranus was discovered in 1781, Neptune in 1846 and Pluto as recently as 1930. The Sun's family is much more extensive than our ancestors used to believe.

The theory that Triton was once independent and that it was subsequently captured by Neptune has been supported by the Voyager 2 results (see Chapter 13). Since 1992 a number of small, icy remote objects have been found in the outer Solar System, in a region known as the "Kuiper Belt", not far beyond the orbit of Neptune, which could also be the dominant source of short-period comets.

2

The discovery of Neptune

The story of the search for Neptune has been told many times, but even today — after the lapse of almost a century and a half — there are still discrepancies in the various accounts, as well as all manner of conflicting opinions. What I propose to do here is to attempt a totally unbiased summary. Whether or not I succeed must be left to the reader to judge!

Seven was the magical number of the ancients. Therefore, it was only natural to assume that there must be seven main bodies in the Solar System: the Sun, the Moon and the five bright planets. The possibility of another planet, moving beyond the orbit of Saturn, was not seriously considered. Then, in 1781, a Hanoverian-born, England-based amateur, William Herschel, was using a home-made telescope to carry out a 'review of the heavens' when he happened upon the world we now call Uranus, after the mythological father of Saturn. Herschel was not making a deliberate hunt for a new planet, and did not immediately recognize it for what it was; his paper to the Royal Society was headed 'An Account of a Comet', but before long its true nature became apparent.

It was found that Uranus fitted in well with a curious relationship known as Bode's Law, which has been described in detail by M. M. Nieto (1972). Because this Law played a major rôle in the later search for Neptune, I must pause to say something about it here. (To be fair, it was not discovered by J. E. Bode, after whom it is usually named, but by an otherwise obscure astronomer, Titius of Wittenberg; Bode merely popularized it during the 1770s.)

Take the numbers 0, 3, 6, 12, 24, 48, 96, 192 and 384, each of which (apart from the first two) is double its predecessor. Add 4 to each. Taking the Earth's distance from the Sun as 10, the distances of the other planets, out as far as Uranus, are represented with tolerable accuracy. This is shown in Table 1; the planets unknown when the Law was first announced are distinguished by asterisks.

It is clear that the agreement is not perfect. All the same, it is fair as far as the bright planets are concerned, and the Law was regarded as being of real significance. There was the problem of the missing planet corresponding to Bode's 28, but, in 1801, Ceres, the largest of the asteroids, was found to fit in quite neatly. So, of

Table 1

Planet	Distance, by Bode's Law	Actual distance
Mercury	4	3.9
Venus	7	7.2
Earth	10	10.0
Mars	16	15.2
*Ceres	28	27.7
Jupiter	52	52.0
Saturn	100	95.4
*Uranus	196	191.8
*Neptune	388	—
*Pluto	—	393.9

course, did Uranus when it was discovered in 1781, and it was logical to think that any more remote planet would correspond to Bode's 388.

In fact it does not. The Law breaks down completely for Neptune, which is the third most massive member of the Sun's family — and when a law fails in such an important case, it ceases to be a Law. The fact that Pluto does correspond reasonably well with Bode's 388 is not in the least significant, both because of Pluto's small size and because its orbit is so eccentric by planetary standards; its 'Bode distance' ranges between 295.8 and 491.9. Moreover, the error even for Saturn is appreciable. Whether or not Bode's Law is more than pure coincidence is still a matter for debate, and there are still some astronomers who retain a degree of faith in it, but it is clearly non-valid at distances beyond that of Uranus — a point which unquestionably misled the subsequent 'Neptune-hunters'.

Though Herschel was the first to realize that Uranus was not a star, he was not the first to record it (and in fact it is clearly visible with the naked eye, under good conditions, if you know where to look for it). Details of pre-discovery observations are given in various books; for instance, in those by A. F. O'D. Alexander (1965), and G. E. Hunt and P. Moore (1988). It was seen by the first Astronomer Royal, John Flamsteed, in 1690, and was given a stellar number: 34 Tauri. Flamsteed also saw Uranus on five other occasions between 1712 and 1715. James Bradley recorded it in 1748, 1750 and 1753; Tobias Mayer saw it in 1756, and the French astronomer Pierre Le Monnier on at least ten occasions between 1764 and 1771, including a run of six observations in January 1769. These old observations were of considerable use later, as we will see.

As soon as the planetary nature of Uranus had been established, its orbit was worked out. By 1802, Pierre Simon de Laplace was able to give its sidereal period as 84.02 years, which is practically correct (the modern value is 84.01 years). But as early as 1788 it had become obvious that Uranus was misbehaving. Mathematicians, including the Rev. Placidus Fixlmillner of Kremsmünster, found that it was straying from its predicted course. It was reasonable to conclude that the old pre-discovery

observations must be in error, so they were discarded; when Fixlmillner did this, he
found that all seemed well.

So it was — but not for long. Within a few years, new discrepancies became
evident. In 1820, the French mathematician Alexis Bouvard decided to reject all the
old observations — a procedure denounced by M. Grosser (1962) as 'shabby' — and
started again, using only the observations made since the discovery of Uranus in
1781. He finally produced new tables (Bouvard 1821).

Even this was not satisfactory, and within four years Uranus was again out of
position. There was something wrong, and astronomers began to think that there
must be an unidentified perturbing influence. One critic of Bouvard's procedure was
Friedrich Wilhelm Bessel, of Königsberg, later to achieve lasting fame as being the
first man to announce the distance of a star (61 Cygni, in 1838, which Bessel
measured by the method of trigonometrical parallax). By 1832, George Airy, then at
Cambridge, reported to the British Association that Bouvard's tables were in error
by half a minute of arc, which was a totally unacceptable amount. It was around this
time that we have the first definite suggestions that an unknown planet might be
responsible.

Airy — later Sir George Airy — plays a dominant rôle in Neptune's discovery, so
it is appropriate to say a little about him here, particularly since his character is all-
important in the whole story. That he was a great astronomer is not in dispute. He
was born in 1801 at Alnwick in Northumberland; he graduated from Cambridge in
1823, and in 1826 was appointed Professor of Astronomy there. In 1835 he was called
to Greenwich to become Astronomer Royal and Director of the Royal Greenwich
Observatory in succession to the Rev. John Pond, whose régime had not been a
success, and who had politely but firmly been requested to resign. Airy was the ideal
choice. He was a first-class organizer, and he restored the Observatory's reputation,
as well as modernizing its equipment and making many other contributions. His
period of office ended only with his retirement in 1881, and he lived on until 1892.

Airy's main defect was an obsession with 'order and method'. The tale that he
once spent a day in the Greenwich cellars labelling empty boxes 'Empty' seems to be
true — and it was also said that to him, the action of wiping a pen on a piece of paper
was as important as the setting-up of a new telescope. He was justifiably confident of
his own ability, but was slow to change his mind even when it had become fairly clear
that he was in the wrong. Neither was he disposed to take advice from others. He
used to insist that the observers remained on duty even when the sky was hopelessly
cloudy; there is a legend that even when rain was falling he used to walk round the
grounds after dark, visiting the telescopes and saying to his observers, 'You are
there, aren't you?' Whether or not this is accurate will never be known, but I
certainly remember being told that Airy's ghost has been seen in modern times
prowling round the Old Royal Observatory on cloudy nights!

After Neptune had been found, Airy gave a detailed account of the circumstances
leading up to the discovery. It was printed in the *Monthly Notices of the Royal
Astronomical Society* for November 1846, and is reproduced in its entirety in
Appendix 2 (Airy 1846). Much of what follows is necessarily taken from this account,
and the details given in this chapter merely summarize what happened.

The possibility of a trans-Uranian planet seems to have occured to several people
at about the same time. One was J. E. B. Valz, Director of the Marseilles

Observatory, who had written to François Arago, France's leading astronomer, about the differences between the theoretical and observed positions of Halley's Comet, which was due to return to perihelion in 1835. Valz believed that the discrepancies could be attributed to a distant planet whose revolution period was at least three times that of the comet. He also mentioned the wanderings of Uranus, and in a letter published later he wrote that 'I would prefer to have recourse to an invisible planet, located beyond Uranus . . . . Would it not be admirable thus to ascertain the existence of a body which we cannot even observe?' (Valz 1835). Another was F. G. B. Nicolai, Director of the Mannheim Observatory, who commented that 'One immediately suspects that a trans-Uranian planet (at a radial distance of 38 astronomical units, according to the well-known rule) may be responsible for this phenomenon' (Nicolai 1836). But the first really definite proposal appears to have come from an amateur, the Rev. T. J. Hussey, Rector of Hayes in Kent, in a letter written to Airy on 17 November 1834. Hussey had had discussions with Alexis Bouvard, and wrote that 'the apparently inexplicable discepancies between the ancient and modern observations suggested to me the possibility of some disturbing body beyond Uranus, not taken into account because unknown'. Apparently Bouvard promised Hussey that he would make some calculations to see whether an approximate position for the planet could be found, but had not in fact done so. Hussey's letter to Airy asked for his opinion as to the feasibility of a search, and he added that with his telescope he thought that he 'could distinguish, almost at once, the differences of light between a small planet and a star'.

Airy's reply, on 23 November, was not encouraging. Even if there were an unknown source of perturbation, he wrote, theory 'was not yet in such a state as to give the smallest hope of making out the nature of any external action' on Uranus. He also felt that Uranus would have to be observed for several successive revolutions before anything of the kind could be contemplated — and since Uranus takes 84 years to complete one journey round the Sun, this would involve a very lengthy delay!

Hussey may have felt rebuffed; in any case he took the matter no further, possibly because he was not in the best of health. The next step came from Eugène Bouvard, Alexis Bouvard's nephew. Eugène discussed the whole problem with his uncle, and in 1837 exchanged letters with Airy, who had by then left Cambridge to take up his post at Greenwich. Again Airy was unimpressed, and wrote that if an unseen body were responsible, 'it will be nearly impossible ever to find out its place'.

As time went by, it became painfully clear that the problem of Uranus would not go away. Bessel, in 1840, gave a lecture in Königsberg in which he commented that with regard to the orbital perturbations, 'Further efforts to account for them must be based on the endeavour to discover an orbit and a mass for some unknown planet, of such a nature that the resulting perturbations of Uranus may be reconciled with the present lack of harmony in the observations'. Bessel also told Sir John Herschel — son of Sir William, the discoverer of Uranus — that in collaboration with his pupil F. W. Flemming he fully intended to take up the investigation (Bessel 1840), but unfortunately this never happened. Flemming died suddenly; Bessel became ill, and he too died in 1846.

Another astronomer of like mind was Johann von Mädler, who, together with Wilhelm Beer, had produced the first really good map and description of the surface of the Moon. The telescope used was Beer's 9.5-cm refractor at his private observatory near Berlin; the map was careful and accurate, and remained the best for half a century. By 1841, Mädler had left Germany to become Director of the new observatory at Dorpat in Estonia, but he retained his interest in the Solar System, and in his book *Populär Astronomie* he wrote: 'We arrive at a planet acting upon and disturbing Uranus; we may express the hope that analysis will at a future time realize in this her highest triumph, a discovery made with the mind's eye, in regions where sight itself is unable to penetrate' (Mädler 1841).

Such was the situation in 1841, when John Couch Adams, a young undergraduate at Cambridge University, made a significant entry in his notebook:

> 1841 July 3. Formed a design, at the beginning of this week, of investigating as soon as possible after taking my degree, the irregularities in the motion of Uranus, which are as yet unaccounted for; in order to find whether they may be attributed to the action of an undiscovered planet beyond it; and if possible thence to determine the elements of its orbit &c approximately, which wd probably lead to its discovery. (Adams 1841)

He did pass his degree — brilliantly — in 1843, and from that time onward the movements of Uranus were very much in this thoughts.

Adams was born at Lidcot, in Cornwall, in 1819, and went to a private school in Devonport before entering Cambridge. He remained at the University after graduating, and in 1860 became Director of the Observatory, where he carried out a great deal of useful research, notably in connection with the Moon's motions and with the orbits of meteor streams. He died in 1892. He was always modest and retiring; he declined the offer of a knighthood from Queen Victoria, and also refused the post of Astronomer Royal when Airy retired in 1881. As a mathematician he had few contemporary equals, and he was well equipped to tackle the problem of Uranus.

By October 1843, Adams had completed most of this research. It was a cosmical detective problem; he could observe the victim (Uranus) and he had to track down the culprit. Undoubtedly he assumed that the planet's mean distance from the Sun must be around 388 on the Bode scale, and this was one reason why his final orbit was incorrect; but at least he was working along the right lines, and, naturally, he communicated with James Challis, Professor of Astronomy at Cambridge. Challis was more receptive than Airy had been to Hussey or Eugène Bouvard, and on 13 February 1844 he wrote to Airy, saying that his 'young friend' Adams was working on the problem and needed some extra information. Airy sent the data required two days later, and Challis passed them on to Adams. In view of what happened later, it is ironical that Airy was so quick to reply to this first inquiry.

By mid-1845 Adams had obtained an approximate position for the new planet, and again Challis wrote to Airy. His letter, dated 22 September, gave all the information, and said that Adams hoped to call to see the Astronomer Royal personally. Adams did so, and it was then that the real chapter of accidents began. Airy was not at home; he had been in France for a meeting. On 21 October Adams called again, not once but twice. On the first occasion Airy was out, and his wife did not know exactly when he would be back, so Adams simply left a card. On the second occasion, the butler told him that Airy was at dinner and could not be disturbed; this may seem odd, since the time was only 3.30 in the afternoon, but Airy had a rigid personal schedule which was never altered. Apparently Airy was not told that Adams had called. Adams did not try again, but left a letter in which he gave the results of his calculations. It is significant that he gave the distance of the unknown planet as 38.4 astronomical units, in conformity with Bode's Law.

On 5 November Airy replied. He expressed interest, but also asked a question about the radius vector which Adams regarded as showing a lack of appreciation about the whole problem. In fact, as A. Chapman has recently stressed, this conclusion of Adams' may have been rather hasty.

The radius vector — that is to say, the imaginary line joining the centre of the Sun to the centre of the planet — is vital in working out any proper tables of motion. Without at least a rough idea of its length, no useful calculations can be made. Since the planet was unknown, it was not possible to do much more than guess at the length of the radius vector. Some sort of an estimate had to be made, and Adams had used Bode's Law as a basis, which under the circumstances was a perfectly reasonable thing to do.

> Though lacking any analytical basis, Bode's Law provided an empirical sequence of numbers which fitted the distances, orbits and radius vectors of the *known* planets. Adams, therefore, began his investigation for the new planet with a theoretical radius vector value based on Bode's Law, from which he could extract the theoretical mass and other elements, by Newton's Laws. The validity of his investigation, however, demanded a leap of faith in the assumption that Bode's Law would also apply to the unknown planet. In Airy's opinion, this was not good enough, and before he, the Astronomer Royal, was willing to devote more time and energy to Adams' ideas, he required physical, not analogical substantiation for this key element of the new planet. (Chapman, 1988).

Adams did not reply to Airy's letter, which proved to be a most unfortunate omission. Later he told a colleague, J. W. L. Glaisher, that he 'should have done so, but the enquiry seemed to me to be trivial'. He was well aware that Airy still regarded the problem as insoluble, and he also knew that the Astronomer Royal was not the sort of man to be challenged by a young and, at that time, unknown astronomer.

Meanwhile, quite unknown to Adams, there had been developments across the Channel. The problem had been taken up by Urbain Jean Joseph Le Verrier, mainly at the instigation of François Arago.

Le Verrier — born at St. Lo, in Normandy, in 1811 — had originally been trained as a chemist, but in 1837 he turned his main attention to mathematical astronomy, and soon earned himself a reputation as a brilliant researcher. Later in his career he became Director of the Paris Observatory. He succeeded Arago there in 1854, but it is fair to say that he had a difficult personality — one of his contemporaries made the acid comment that although he may not have been the most detestable man in France, he was certainly the most detested — and in 1870 he was dismissed as Director on the grounds of his 'irritability'. However, he was reinstated when his successor, Charles Delaunay, was drowned in a boating accident, and retained the post until his death in 1877. Le Verrier's reputation for awkwardness has persisted to this day, though in view of his cordial relationship with Adams it may be that history has treated him rather unkindly.

Le Verrier was faced with exactly the same problem as Adams. His method was different in approach, but the final result was almost the same. In 1845 he published a memoir on the subject in the *Comptes rendus* of the Paris Academy of Sciences (Le Verrier 1845) and in December of that year it arrived on Airy's desk. In the following summer came another memoir (Le Verrier 1846a) and Airy received it in late June. Airy was impressed, and even wrote that 'I cannot sufficiently express the feeling of delight and satisfaction which I received from it'. Le Verrier's position for the new planet agreed with Adams' to within almost one degree of arc, and Airy was stirred into action.

Within twenty-four hours a letter from Airy was on its way to Le Verrier, though, perhaps significantly, it contained no mention of Adams. In it, Airy asked the same question about the radius vector that he had put to Adams. Probably Le Verrier, too, regarded it as 'trivial', but he lost no time in sending a satisfactory answer; his letter was clear, concise and convincing. Airy's last doubts vanished, and he realized that a prompt search for the new planet was essential if the prize was not to slip through England's grasp. Yet there was still virtually no direct contact between Airy and Adams; on 2 July they met by chance while they were walking across St. John's Bridge in Cambridge, but their conversation lasted for no more than a couple of minutes, and apparently the Uranus problem was mentioned either briefly or not at all.

Peter Hansen, Director of the Seeberg Observatory in Denmark, had been staying with Airy at Greenwich and was with him at the time of the chance meeting, but whether Airy told him anything about Adams is doubtful. He had, however, discussed the problem earlier with James Challis and John Herschel, and had referred to 'the extreme probability of now discovering a new planet in a very short time, provided the powers of one observatory could be directed to search for it'. Herschel was enthusiastic. Surprisingly, he was more so than Challis, whose relationship with Airy was not entirely harmonious.

Since Airy was Director of the country's leading observatory, it might be thought that he would personally organize a search from Greenwich. This is precisely what he did not do. Quite apart from the fact that it would mean disrupting the routine work (a prospect which Airy viewed with extreme disfavour), there was no large telescope available there. The most powerful instrument was the Sheepshanks refractor, with an aperture of a mere 17 centimetres. At Cambridge, Challis had the Northumberland refractor, with its 29.8-cm object-glass, which was why Airy decided to leave things in Challis' hands. On 9 July he wrote to Challis, stressing the importance of the search and the need for urgency. He followed this up with a second letter on 13 July, telling Challis how the search should be conducted and offering the help of an assistant from Greenwich if need be.

Challis had been away. On his return, on 18 July, he wrote to Airy, agreeing to do as he had been asked but declining the offer of an extra helper. Even then, he did not act swiftly; his first telescopic sweeps were not made until 29 July. He had no really detailed map of the area of the sky in which the planet was thought to lie, and he adopted the cumbersome method of using a magnifying power of 166 (giving a field of 9 minutes of arc) and checking on all the stars which came successively into view during the sweeps. Adams had estimated that the planet would be at least as bright as magnitude 9 (actually, of course, it is above 8). Challis was not so optimistic, and decided to map all the stars down to magnitude 11. Since there were several thousand stars of this brightness in the relevant region of the sky, it was clear from the outset that the search would be a protracted one. During his observations, he even wrote to Airy — who was temporarily away in the Continent — saying that although had had been favoured with good weather, and had taken a considerable number of observations, 'I get over the ground very slowly . . . . I find that to scrutinize thoroughly, in this way, the proposed portion of the heavens, will require many more observations than I can take this year'.

A more skilful and clear-thinking observer would no doubt have acted differently, but Challis was not of this calibre. He had been born at Braintree in Essex in 1803, educated at a local school and then sent to Cambridge University. Academically his ability left nothing to be desired; he graduated with top honours in 1825, after which he was ordained and served as Rector of a Cambridgeshire parish from 1830 to 1832 before becoming Plumian Professor of Astronomy at the University in 1836. He also acted as Director of the Observatory between 1836 and 1861, and he carried out some useful work. Unfortunately, he was indecisive and lacking in self-confidence; this irked Airy, who did not have much patience with those of weaker character than himself, and relations between the two became somewhat strained. True, Challis conducted the search for Neptune very much as Airy had recommended, but it was not in his nature to strike out for himself, and by this time his faith in Adams' calculations had waned alarmingly.

He plodded on, as we learn from his own account delivered to the Royal Astronomical Society on 13 November 1846 immediately after Airy's (Challis 1846a). But by then, quite unknown to him, he was no longer alone in his quest.

Le Verrier's next memoir was presented to the Académie des Sciences on 31 August (Le Verrier 1846b). In it he gave the orbital elements — again with a distance which corresponded more or less with Bode's Law — and stated that the planet would be found about five degrees east of the third-magnitude star δ Capricorni. He also claimed that the apparent diameter would be 3″.3 or thereabouts, so that even a moderate telescope would show the planet as a small disk rather than as a starlike point. Le Verrier himself was not a practical observer, but undoubtedly he expected co-operation from his colleagues at the Paris Observatory, particularly as he had started his mathematical research at the request of Arago himself.

He was soon disillusioned. The French observers were as lack-lustre as Airy had originally been. By 18 September his patience — never his strongest virtue — had snapped. He had had some connections with Johann Gottfried Galle, of the Berlin Observatory, and he therefore wrote to Galle asking him to begin searching in the position indicated (Le Verrier 1846e; surprisingly, it seems that the letter was not published in full until 1910). Le Verrier gave the distance of the planet as 36.154 astronomical units, which is 361.5 on the Bode scale, and the revolution period as 217.387 years.

Galle sought, and obtained, the permission of the Director of the Observatory, Johann Encke (best remembered today for his work in connection with the periodical comet which bears his name, and also for his detection of the division in Saturn's Ring A). During their talk, Galle and Encke were joined by a young student, Heinrich Louis D'Arrest, who promptly asked to be allowed to join in. Galle, feeling that it would have been 'unkind to refuse the wish of this zealous young astronomer', agreed, and on the same night they began work, using the fine 23-cm Fraunhofer refractor, which was probably the best telescope of the time.

Galle was no Challis. He had instinctive faith in Le Verrier's calculations, and he began by turning the telescope to the position which he had been given: R.A. 22h 46 m, declination −13° 24′. He hoped to find an object which showed a disk. Meanwhile, D'Arrest suggested that it would be a good idea to use a star map. By good fortune there was one to hand, *Hora XXI* of the Berlin Academy's Star Atlas, drawn up by Bremiker, which had already been printed but had not yet been widely distributed to other observatories.

They settled back to work. Galle, at the telescope, called out the positions and appearances of the stars which came into view; D'Arrest checked them against the chart. They did not have long to wait. Within minutes Galle described an 8th-magnitude star at R.A. 22 h 53 m 25 s.84. D'Arrest called out: 'That star is not on the map!'

Probably the two realized at once that the hunt was over, but it was essential to be sure, particularly as the disk was so small that the object could quite easily have been mistaken for a star. Encke joined them, and they tracked the object until it set. On the next night, 24 September, they found that it had moved, and also that it really did show a measurable disk; Encke made the diameter 3″.2, in almost perfect agreement with Le Verrier's predicted 3″.3. On the following day, Galle wrote the letter to Le Verrier quoted at the beginning of this book. 'The planet whose position you have pointed out actually exists' (Galle 1846). And on 28 September Encke wrote glowingly to Le Verrier:

Allow me, Sir, to congratulate you most sincerely on the brilliant discovery with which you have enriched astronomy. Your name will be forever linked with the most outstanding conceivable proof of the validity of universal gravitation, and I believe that these few words sum up all that the ambition of a scientist can wish for it. It would be superflous to add anything more . (Encke 1846b)

Encke also wrote to Heinrich Schumacher, editor of the periodical *Astonomische Nachrichten*, and to Hansen in Denmark. Airy had just arrived at Hansen's observatory, and it was there that he learned of the discovery, several days before it percolated as far as England. Airy, however, did nothing, and it was not for a fortnight that he made his first comments. It is possible that he was already aware that he was bound to be criticized for his inertia.

Challis was still in blissful ignorance of events on the Continent. He was following Airy's instructions, but without any real sense of urgency, and he was also preoccupied with observation of Biela's periodical comet, which had caused an astronomical sensation of the first magnitude by splitting into two pieces. (The twins returned on schedule in 1852, but have never been seen since, and there is no doubt that Biela's Comet is dead; for years it produced meteors radiating from the constellation of Andromeda, but even the shower has now virtually ceased.) It was only on 29 September that Challis saw Le Verrier's memoir of 31 August. In this memoir, Le

Verrier had claimed that the planet would show a perceptible disk. Challis at once decided to look — and then followed another example of almost incredible incompetence.

From all accounts, Challis surveyed over three hundred stars on the night of 29 September, and one of them attracted his attention; he told his assistant to add the note 'It seems to have a disk'. Remember, he was using a magnification of only 116, and the Northumberland refractor would bear a power much higher than that, but Challis did not take the trouble to see whether the 'disk' would show up more clearly with a more powerful eyepiece. Evidently he meant to do so on the following night, but clouds intervened, and by 30 September the Moon had become unpleasantly obtrusive, though there is no reason to suppose that it would have prevented the observations if Challis had been really enthusiastic about making them. The chance had been lost. On 1 October a letter from John Russell Hind to the London *Times* showed Challis that he had been forestalled:

> I have just received a letter from Dr. Brünnow, of the Royal Observatory at Berlin, giving the very important information that Le Verrier's planet was found by Mr. Galle on the night of September 23 . . . . This discovery may be justly considered one of the greatest triumphs of theoretical astronomy. (Hind 1846)

Hind, formerly an assistant at Greenwich, had become Director of George Bishop's private observatory in Regent's Park (of which, alas, nothing now remains), and had the use of a 17.7-cm refractor, larger than the Greenwich Sheepshanks but much smaller than the Cambridge Northumberland. Hind had presumably received Brünnow's letter before the end of September, because on the 30th, despite the moonlight, he located the new planet and saw that it did indeed show a perceptable disk.

Even Challis must have felt humiliated, but there was worse to come. When he checked back on the series of measurements he had started on 29 July, he found that he had recorded the planet twice during the first four nights of observing, on 30 July and 4 August. It was again recorded on 12 August, and, of course, it was also the object which he had suspected of showing a disk. If only he had compared his observations, he could not have failed to identify the planet. He admitted, rather lamely, that

> his oversight was partly caused by the pressure of comet reductions, but principally from an impression that a long search was required to ensure success. He was also anxious to secure the greatest number of observations, and so postponed the comparison till he had greater leisure. He admits, moreover, that he had too little confidence in the indications of theory, though perhaps as much as others might have felt in similar circumstances, and with similar engagements. (Challis 1846a)

There is yet another episode which casts further light on Challis's attitude. Unfortunately we do not have a first-hand report of it; it was given in an account written a century later by W. M. Smart (Smart 1947, p. 63) but there seems no reason to doubt it. It seems that the Rev. William Kingsley, of Sidney Sussex College in

Cambridge, was dining with Challis in Trinity College when Challis told him about the star which seemed to have a disk. Kingsley suggested using a higher magnification, and Challis answered: 'Yes, if you will come up with me when dinner is over we will have a look at it.' They went to Challis' rooms in the Observatory, and the sky was clear, but they delayed because Mrs. Challis invited them to have a cup of tea — and by the time that they reached the dome of the Northumberland telescope, clouds had come up. Kingsley does not give the precise date (or, at least, it is not in Smart's account), but it must presumably have been 29 September.

There is no doubt that Adams was the first to give a reasonably accurate position for Neptune, and if a search had been organized without delay he would have had the unquestioned honour of priority. And in this connection it is worth relating a story which, if true, would have been yet another to add to the chapter of accidents. It was first told by E. S. Holden, then Director of the Lick Observatory, many years after the actual events (Holden 1892) and it has been repeated many times, even by M. Grosser in his detailed book *The Discovery of Neptune* (Grosser 1962, p. 92).

According to Holden, Airy had a conversation with a well-known English amateur astronomer, the Rev. William Rutter Dawes, shortly after he received Adams' results in October 1845. Dawes at once wrote to another amateur, William Lassell, who was by profession a brewer, but who had become an expert observer and was just setting up a 61-cm reflector in his private observatory near Liverpool. Dawes 'begged' Lassell to search for the new planet, and gave him all the data. Unfortunately Lassell was in bed with a sprained ankle; he put Dawes' letter aside for attention as soon as possible, but a careless housemaid 'tidied it up', and Lassell took no further action.

Holden visited England in 1876 and saw Lassell, who was then aged 77. The story was actually related to Holden by Lassell's wife, but it had never been made public, and Holden kept to to himself until both Lassell and Adams had died.

All this may sound plausible, but there are strong reasons for believing that it is untrue, as has been pointed out by R. W. Smith in a recent paper (Smith 1983). As Smith rightly says, Holden did not hear the story until a full thirty years after the events are supposed to have taken place. It was only in September 1846 that Adams sent Airy a full account of his work, so that Airy could not have mentioned it to Dawes in 1845. We know that in February 1846 Lassell visited Dawes at his home in Kent, so that he could have obtained a copy of the 'lost' letter. Letters between Airy, Challis and Lassell in December 1845, now preserved in the Cambridge Observatory archives, give no hint that Lassell knew anything about the quest for a new planet, and neither is it mentioned in Lassell's surviving notebooks of 1846. Moreover, when Lassell heard about the discovery and looked at Neptune with his 61-cm reflector, on 2 October, he saw the disk at once (Lassell 1846), so that if he had searched in October 1845 he could not have overlooked it. Finally, and most significant of all, Lassell was not the sort of man to ignore a challenge, sprained ankle or no sprained ankle — and it was also he who discovered Triton, Neptune's larger satellite, soon after the planet itself had been identified. All in all, there seems no doubt that Smith is correct in discounting the whole story.

Then who was to blame? Airy received the most criticism, as we will see. According to W. Ellis, who knew him, Airy was 'abused most savagely', and the whole affair 'seemed unduly to overshadow him for the rest of his life' (Ellis 1905). A hundred years after the discovery, at an address to the Royal Astronomical Society on the occasion of the centenary celebrations, Airy's reputation also came under fire from W. M. Smart, who commented that his treatment of Adams in general was 'unbecoming to the leading astronomer of his generation' (Smart 1947, p. 650). A different view was taken by the then Astronomer Royal, Sir Harold Spencer Jones, who stressed the extent of Airy's work load at the time, and described Smart's strictures as 'unjustifiable' (Jones 1947). Even after so long an interval, feelings still ran high.

Perhaps the last word on the subject has been said by Chapman in his recent paper. Airy's rôle as Astronomer Royal was rather different from that of subsequent holders of the office. Quite apart from astronomy, he had public responsibilities in a wide range of activities, and around 1846 he was preoccupied with the Railway Gauge Commission, which probably took more of his time than anything else during this particular period. He was very conscious of his public duties — after all, he was paid by the State, and unlike most of his predecessors he had no significant private income — and to Airy, his official duties were paramount. Therefore, he was not inclined to indulge in what might be termed, in modern parlance, wild goose chases. In this respect Chapman maintains that he was more sinned against than sinning. 'While it is futile to speculate what might have happened had certain persons acted differently, what cannot be denied is that the search for Neptune was the first major incident to highlight the distinction between a scientist's private research and what was seen to be the state's public duty' (Chapman 1988).

Even if Airy had been dilatory, Challis, in the end, was even more so. In 1963 the well-known astronomical historian H. H. Turner, himself a Cambridge graduate, wrote:

> I have always felt that my old University made a scapegoat of the wrong man in venting its fury upon Airy, when the real culprit was among themselves, and was the man they had themselves chosen to represent astronomy. He was presumably the best they had; but if they had no one better than this, they should not have been surprised, and must not complain, if things went wrong. (Turner 1963, p. 71)

Turner also relates that Challis was used to 'leaning rather helplessly upon Airy', and that on one occasion he even implored the Astronomer Royal to go down to Cambridge and show how to lecture to students — which Airy duly did.

When the discovery was made, Le Verrier, naturally, was completely unaware that anyone else had been making similar calculations about the movements of Uranus: remember that Adams had published nothing. The first announcement was made by Sir John Herschel in the London *Athenæum* on 3 October following the discovery. Herschel wrote as follows:

> The remarkable calculations of M. Le Verrier — which have pointed out, as now appears, nearby the true situation of the new planet, by resolving the inverse problem of the perturbations — if uncorroborated by repetition of the numerical calculations by another hand, or by independent investigation from another quarter, would hardly justify so strong an assurance as that conveyed

by my expression above alluded to. But it was known to me, at that time, (I will take the liberty to cite the Astronomer-Royal as my authority), that a similar investigation had been entered into, and a calculation as to the situation of the new planet very nearly coincident with Mr. Le Verrier's arrived at (in entire ignorance of his conclusions), by a young Cambridge mathematician, Mr. Adams; — who will, I hope, pardon this mention of his name (the matter being one of great historical moment), — and who will, doubtless, in his own good time and manner, place his calculations before the public. (Herschel 1846)

The storm was gathering — and the situation was not improved by two unfortunate letters, one written by Challis and the other by Airy. On 5 October Challis wrote to Arago, saying that he had been engaged in a search for a new planet; that he had detected an object which seemed to show a disk, and that he was able to confirm, 'in a remarkable manner, the accuracy of the conclusion which M. Le Verrier arrived at from theoretical conclusions, that with a good telescope the planet could be distinguished by its physical appearance' (Challis 1846b). Challis did not mention Adams at all, but Airy evidently felt that he had no choice. On 14 October he wrote to Le Verrier in a rather transparent effort to pour oil on potentially troubled waters. After fulsome congratulations to Le Verrier, he went on:

I do not know whether you are aware that collateral researches had been going on in England and that they led to precisely the same results as yours. I think it probable that I shall be called on to give an account of these. If in this I shall give praise to others, I beg that you will not consider it as at all interfering with my acknowledgement of your claims. You are to be recognized beyond doubt as the real predictor of the planet's place. I may add that the English investigations, as I believe, were not quite so extensive as yours. They were known to me earlier than yours.

Yet on the same day he wrote to Challis: 'Heartily do I wish that you had picked up the planet, I mean in the eyes of the public, because in my eyes you have done so. But these misses are sometimes nearly unavoidable.' And in a later letter to J. B. Biot, in June 1847, he wrote: 'I assure you that I have a very high opinion of Mr. Adams and that upon the whole I think his mathematical investigations superior to M. Le Verrier's.' Airy can hardly lay claim to consistency!

When the French realized that Le Verrier's claim to absolute priority was being challenged, they were furious. François Arago led the onslaught. He promptly published a thunderous and emotional attack:

What! M. Le Verrier has made his researches available to the entire scientific world; following the formulæ of our learned compatriot, everyone was able to see the new planet dawn, rapidly become brighter, and before long appear in its full brilliance; and today we are called upon to share this glory, acquired so diligently and legitimately, with a young man who has communicated nothing to the public, and whose calculations, more or less incomplete, are with only two exceptions totally unknown to the observatories of Europe! No, no! The friends of science will not permit the perpetration of such a flagrant injustice! The journals, the letters which I have received from several English scientists prove to me that in that country, too, the eminently respectable rights of our compatriot will find zealous defenders.

Mr. Adams has no right to figure in the history of the discovery of the planet Le Verrier, neither by a detailed citation, nor by the slightest allusion. (Arago 1846a)

This was only the start. Further attacks and counter-attacks followed, and it has been claimed that the affair almost led to an international incident. A well-known British astronomer, J. P. Nichol, wrote: 'Let me deplore the conduct of M. Arago. ... I lament the following explosion of Gallic intolerance, which I fear for ever incapacities him for the office of judge' (Nichol 1849). R. Grant described Arago's attitude as 'monstrous' (Grant 1852). In France, the popular press joined in; for instance *L'Univers* (21 October 1846), *Le National* (also 21 October) and *L'Illustration* (a series of unpleasant caricatures of Adams, 7 November). The efforts of Airy and Challis to calm the situation seem, initially at least, to have met with scant success.

Challis evidently had some sort of struggle with his conscience, as is shown for his report to the Cambridge Observatory in December 1846. After describing the course of events, he went on:

My success might have been more complete, if I had trusted more implicitly the indications of the Theory. It must, however, be remembered that I was in quite a novel position. The history of Astronomy does not afford a parallel instance of observations undertaken entirely in reliance upon deductions from theoretical calculations, and those too of a kind before untried .... And that all this may be said, is entirely due to the talents and labours of one individual among us, who has at once done honour to the University, and maintained the scientific reputation of the country. It is regretted that Mr. Adams was more intent upon bringing his calculations to perfection, than on establishing his claims to priority by early publication. Some may be of the opinion, that in placing before the first Astronomer of the kingdom, results which shewed that he had completed the solution to the Problem, and by which he was, in a manner, pledged to the publication of his calculations, there was as much publication as was justifiable on the part of a mathematician whose name was not yet before the world, the theory being one by which it was possible that the practical astronomer might be misled. Now that success has attended a different course, this will probably not be the general opinion. I should consider myself to be hardly doing justice to Mr. Adams, if I did not take this opportunity of stating, from the means I have had of judging, that it was impossible for anyone to have comprehended more fully and clearly all the parts of this intricate Problem; that he carefully considered all that was necessary for its exact solution; and that he had a firm conviction, from the results of his calculations, that a Planet was to be found. (Challis 1846c)

It is hard to credit that this was the same man who had earlier written a congratulatory letter to Arago without mentioning Adams at all.

There was also the curious affair of the British medals. The Royal Astronomical

Society debated whether or not a Medal should be awarded in the year 1846; the first name considered was Le Verrier's, but it was thought that 'an award to M. Le Verrier, unaccompanied by another to Mr. Adams, would be drawing a greater distinction between the two than fairly represents the proper inference from facts, and would be an injustice to the latter'. Since the rules of the Society prohibited the award of more than one medal per year, it was thought prudent to award no medal at all (Royal Astronomical Society report, 1847). The Royal Society of London acted differently when they came to consider their most prestigious award, the Copley Medal. On 30 November 1746 they made the award — to Le Verrier. This led to criticism; for instance, J. P. Nichol went so far as to speak of British men of science 'bowing before Gallic pretension' (Nichol 1849).

The whole situation was regrettable, but it would have become ten times worse if either Adams or Le Verrier had taken part. To their eternal credit, they did not do so. It was clear to all fair-minded people that they were entitled to rank as co-discoverers, and when they met for the first time, at the meeting of the British Association in June 1847, they struck up an immediate friendship which lasted for the rest of their lives. John Herschel, in an address to the Royal Astronomical Society in 1848, said that 'there is not, nor henceforth ever can be, the slightest rivalry on the subject between these two illustrious men — as they have met as brothers and as such will, I trust, ever regard each other — we have made, we could not make, no distinction between them on this occasion' (Herschel 1848). There is also a charming account of Adams by a family friend, Caroline Fox, dating from 1847:

Adams is a quiet-looking man, with a broad forehead, a mild face and most amiable and expressive mouth. I sat by him at dinner, and by gradual and dainty approaches began to get at the subject on which one most wished to hear him talk. He began very blushingly, but went on to talk in a most delightful fashion, with large and luminous simplicity, of some of the vast Mathematical facts of which he is so conversant . . . He speaks with the warmest admiration of Le Verrier, especially of his exhaustive method of making out the orbits of comets, imaging and disproving all tracks but the right one — a work of infinite labour . . . . They enjoyed being a good deal together at the British Association meeting at Oxford, though it was unfortunate for the intercourse of the fellow workers that one could not speak French nor the other English! He had met very little mathematical sympathy from Challis, of the Cambridge Observatory, but when his result was announced, there was noise and enough to spare. (Troy 1969)

There had, however, been an earlier controversy over the name for the new planet, and in this Le Verrier did play a part.

When, in 1781, Uranus had been discovered, there had been some discussion over an official name. Its discoverer, William Herschel, wanted to call it '*Georgium Sidus*' in honour of his patron, King George III of England and Hanover, but this was never popular, even though many people referred to the planet as either 'The Georgian' or even 'Herschel' until well into the nineteenth century. However, Johann Elert Bode's suggestion of 'Uranus' met with general approval. In mythology, Uranus was the father of Saturn, and since all the other planets had mythological names it seemed sensible to carry on with the tradition.

When the new planet was found, in 1846, Challis suggested the name 'Oceanus'. Galle, in a letter to Le Verrier, preferred 'Janus'. But when Le Verrier wrote to Galle, on 1 October 1846, he said: 'The Bureau of Longitudes here has decided upon "Neptune". The symbol is to be a trident. The name Janus would imply that this is the last planet in the Solar System, which we have no reason at all to believe.'

In fact, it does not seem that the Bureau of Longitudes made any recommendation at all, and the name 'Neptune' was more probably Le Verrier's own suggestion. Yet between 1 October and 5 October he altered his mind, and decided that he wanted the planet to be named after himself. Arago fell in with this suggestion, and at the meeting of the French Académie des Sciences on 5 October he announced that since Le Verrier had delegated to him the choice of name, he had decided 'to call it by the name of the man who so learnedly discovered it' (Arago 1846a, p. 662). He went on: 'I pledge myself never to call the new planet by any name other than Le Verrier's Planet. I believe I will thus give an irrefutable proof of my love for science, and follow the inspiration of a legitimate patriotism.' He printed this decision on 5 November (Arago 1846b). He added that in future he proposed to call the seventh planet 'Herschel' rather than 'Uranus'.

Arago's proposal was not well received. For example, Encke, from Berlin, said that he found 'the name Neptune, chosen by Le Verrier, perfectly appropriate'. Sir Harold Spencer Jones, in his centenary booklet, cited a significant comment made to Airy by Admiral W. H. Smyth, the distinguished English amateur astronomer: 'Mythology is neutral ground. Herschel is a good name enough. Le Verrier somehow or other suggests a Fabriquant and is therefore not so good. But just think how awkward it would be if the next planet should be discovered by a German, by a Bugge, a Funk, or your hirsute friend Boguslawski?' (Jones 1946).

The argument rumbled on for some time, but there was never any serious doubt that 'Neptune' would prevail, as indeed it did (Encke, 1846b). Oddly enough, Uranus was still referred to as 'the Georgian' in the British Nautical Almanac until as late as 1850, by which time the name of Neptune for the other giant had become firmly established.

One question remains to be answered. If Adams was so sure that he had at least an approximate position for Neptune, then why did he not go and look for it? His estimated magnitude was about 8, which is close to the truth, and there seems to be no obvious reason why he delegated the search to others.

Thanks to his descendant Mrs Norma Foster, I have been able to read some letters which may throw at least a little light on the subject. Like many other theorists, he was simply unused to observing, and whether he had ever used the Northumberland refractor seems debatable. However, in 1844 he wrote to his younger brother George (later Mayor of Saltash) as follows, evidently after an earlier request that a telescope should be sent direct to him:

My dear Brother,
    I should have written to you long before this had I not been so very busy this term to have the time for letters. I am much obliged to you for sending the telescope so carefully. I have not used this, as the weather has not been possibly

good enough to try it so much as I could wish, but I think it is a very good glass. O'Reilly, a friend of mine, who has been at sea, has lent me his telescope, which he tells me is the best of the kind he has ever met with in the fleet, I hope we shall be able to give both a fair trial Xmas . . . I don't know whether I ever mentioned that we have a small observatory attached to the college. Mr. Griffin and I have just been appointed joint Curators of it, so I think I now have the opportunity of using a pretty good telescope and several other astronomical instruments. (Adams 18 December 1844)

But it is clear that he made no personal hunt for Neptune — and it is ironical that he would not even have needed a good telescope. It has been a question which has puzzled me for years, and some time ago (in 1982) I decided to put it to the test.

I did not know the precise position of Neptune, and I carefully refrained from looking it up. I asked a colleague to give me an 'error circle' around the actual position which was the same as the probable error of Adams' calculated position. I then went out, armed with a pair of 10×50 binoculars, and began to look. I knew that with this magnification Neptune would look exactly like a star, and so I began to plot all the stars above magnitude $8\frac{1}{2}$ within my 'error box', hoping that I would able to identify Neptune by its movement. I did so; it took me 17 nights. What I could do in 1982, Adams could have done in 1845 — and in this case he would have been the sole claimant to both the correct prediction and the actual discovery of Neptune.

To check further on Challis' observations, I used the 33-cm and 66-cm refractors at the Royal Greenwich Observatory, Herstmonceux in September 1988, to look at Neptune. With a power of $\times 110$, there was absolutely no chance of confusing Neptune with a star; it showed a distant disk, and was clearly bluish. Triton was very easy. With a higher power of $\times 250$, the distinction between Neptune and the star was even more obvious.

Even now we have not heard the last of the argument. As I hope I have shown, it is wrong to assume that Airy and Challis are the black villains and Adams and Le Verrier the dauntless heroes; matters are much more complicated than that. Everyone is entitled to an opinion, and some will no doubt disagree with the way in which I have presented the story. At least it would be both sad and unfair if Airy, in particular, were to be remembered not as the great astronomer and administrator which he undoubtedly was, but as the man who lost the chance of claiming Neptune as a British discovery when all the clues had been delivered into his hands.

A new biography of Adams, *Voyager in Space and Time*, has been published by the Book Guild (Lewes, East Sussex). It has been written by Adams' great-great niece, Hilda Mary Harrison, and contains much new material, though it does not in any way alter the arguments put forward above.

3

Pre-discovery observations: and the orbit of Neptune

After the discovery of Uranus, in 1781, it was found that the planet had been recorded on quite a number of occasions and had been mistaken for a star; as we have noted, Flamsteed even gave it a number — 34 Tauri. Since Neptune is brighter than the 8th magnitude, there was every reason to suspect that it, too, might have been seen several times. Pre-discovery observations were important in the determination of the orbit; after all, Neptune is a slow mover, and takes well over a century and a half to complete one orbit round the Sun. Galle first identified it less than one 'Neptunian year' ago.

Apparently the first astronomer to check for pre-discovery observations was S. C. Walker, who had, incidentally, planned to search for Neptune at the United States Naval Observatory, but had been unable to get started before Galle's announcement. Walker examined the star catalogue compiled late in the eighteenth century by the famous French astronomer J. J. de Lalande (Lalande 1801) and soon found what he had been seeking. Lalande had observed the region of Neptune on 8 May 1795 and again on 10 May, and had recorded an 8th-magnitude star within two minutes of arc of where Neptune had actually been. Walker accordingly looked at the area, and, as he expected, Lalande's 'star' was missing (Walker 1847a). Using Lalande's position, Walker promptly made new calculations to improve the published orbital elements of Neptune, which, understandably, were then poorly known.

The Lalande observations were also studied by A. C. Petersen, assistant to H. Schumacher, the founder of the periodical *Astronomische Nachrichten*. Petersen agreed with Walker's findings (Petersen 1847), but also noted that Lalande had followed his observation with the symbol ':', meaning 'doubtful position'. Next, F. Mauvais referred to Lalande's original manuscript, and deduced that both the 8 May and 10 May observations were of Neptune — but because they differed in position, Lalande had rejected the first entirely and had questioned the second (Mauvais 1847). It is fair to say, then, that Lalande really did have Neptune in his grip.

Other pre-discovery observations had been made by the Scottish astronomer

John Lamont, sometimes referred to as Johann von Lamont because he spent most of his career in Germany (he was appointed Director of the Munich Observatory in 1835). Lamont had seen Neptune three times: on 25 October 1845, and on 7 and 11 September 1846 (Lamont 1850, 1851). Like Lalande, he had failed to realize that it was anything but a star. His observations were later reduced by John Russell Hind, who had been one of the first to confirm the discovery of Neptune. Hind gave Lamont's position for the planet as R.A. 21 h 42.m 42 s.48, north polar distance 104°14'23".0, and R.A. 21 h 54 m 44 s.1, north polar distance 103°16'21".8 (Hind 1850). As he pointed out, Lamont would have identified Neptune if he had been prompt to compare the observations of 7 and 11 September (Hind 1851).

Yet there was an even more remarkable pre-discovery observation which was not appreciated until over a hundred years after the identification of Neptune. The first recorded observation was made in 1612 by no less a person than Galileo!

In that year Galileo, using one of his early telescopes, was busily observing Jupiter and its four bright satellites. S. C. Albers has calculated that Jupiter actually occulted Neptune in 1613, so that throughout the whole of that period the two planets were close together (Albers 1979). It therefore occurred to Charles T. Kowal and Stillman Drake to look carefully at Galileo's observations, and they found something very interesting indeed (Kowal and Drake 1980).

Galileo made a sketch dated 27 December 1612, at 15.46 a.m.; since he considered each day to begin at noon, this corresponded to 28 December, 03.46 local time. A 'star' was shown corresponding to the position of Neptune. The distance was wrong, as Neptune would have been off the edge of the page of the notebook, so Galileo put it in at the edge and indicated its direction with a dotted line.

On 2 January 1614 Galileo recorded the 7.1-magnitude star SAO 119234 near Jupiter, and followed the same procedure. The star was again shown on 28 January and was marked 'a', while the other object, 'b', must be Neptune. Galileo's note in the lower left quarter of his drawing reads: 'Beyond fixed star a, another followed in the same straight line, which was also observed on the preceding night, but they then seemed further apart from one another.' In fact, Galileo had detected Neptune's motion. The telescope used seems to have had a magnification of ×18 and a resolving power of 10 seconds of arc, with a field of view 17 minutes of arc in diameter. Neptune's magnitude was 7.7, and Galileo often plotted stars fainter than that.

D. Hughes has stated that since Jupiter and Neptune were then in Virgo, 'two factors can account for Galileo not noticing the movement of Neptune against the fixed star background. First, there were no bright stars near. Second, the mean daily motion is only 22 seconds of arc — too small for him to detect. So it is wrong to assume that Neptune was within his grasp' (Hughes 1980). Actually Galileo *did* detect the motion, but it is not surprising that he failed to realize its significance. D. Rawlins and E. M. Standish (1981) have questioned Galileo's drawing and its scale, but at least there is little doubt that the star 'b' really is Neptune. It would have been ironical if Neptune had been identified almost a hundred and seventy years before the much closer and brighter Uranus!

From all this, Kowal and Drake concluded that the orbital elements of Neptune were in need of revision, even though Galileo's observation of 28 January 1613 seems to have been within one minute of arc of the position according to present-day theory. Kowal and Drake also referred to the fact that Lalande's pre-discovery position of 1795 differed from theory by as much as 7 seconds of arc, though this had been noted earlier by Rawlins (1970) who suggested that it might be due to an error in observation.

Almost as soon as Neptune had been identified, one fact became apparent: Bode's Law had broken down. Encke, for example, gave the 'Bode distance' as 300 instead of the predicted 388 — that is to say, a real distance of 30.038 85 astronomical units (Encke 1847). Moreover, even though Adams and Le Verrier had been so accurate in predicting Neptune's position, their orbital elements were badly wrong. In 1847 Walker worked out an orbit from the new observations (Walker 1847b) and, as can be seen from Table 2, the discrepancies were very marked indeed.

Table 2

	Adams	Le Verrier	Walker	Modern
Mean distance from Sun (astronomical units)	37.25	36.15	20.35	30.06
Orbital eccentricity	0.120 62	0.107 61	0.008 84	0.008 86
Sidereal period (years)	227.3	217.387	166.381	164.788
Mass (Sun=1)	1/6666	1/9300	1/15 000	1/19 300
Longitude of perihelion	299°11′	284°45′	0°12′25″	. . .
True longitude, 1 Jan. 1847	329°57′	326°32′	328°7′34″	. . .

Bode's law had proved to be a cosmical red herring. Professor Benjamin Peirce, of Harvard, was quick to declare that Neptune was not 'Le Verrier's Planet' at all. Basing his conclusions upon observations made by W. C. Bond at Harvard (Bond 1847) and by himself, Peirce wrote that

the planet Neptune is not the planet to which geometrical analysis has directed the telescope; that its orbit is not contained within the limits of space which have been explored by geometers searching for the source of the disturbances of Uranus; and that its discovery by Galle must be regarded as a happy accident. (Peirce 1847 6a).

Peirce followed his attack with a second paper (Peirce 1847b) in which he used a mass for Neptune of 1/19 840, which is much closer to the true value than any of the previous estimates. He sent his results to Walker, who produced an improved orbit. Predictably, other astronomers were not impressed with the 'happy accident' theory, and it is worth quoting some comments made in 1850 by Benjamin Apthorp Gould, of Boston, who wrote an account of the history of Neptune's discovery (Gould 1850):

The arguments which tend to prove that Neptune is the planet of their theory can only be based upon the supposition of error in that theory, a supposition which I am unwilling to admit. Investigations conducted with the care and precision which characterized these must not be so lightly dealt with. The combined labours of Le Verrier and Peirce have incontrovertibly proved, that, by reducing the limits of error assumed by the modern observations to 3″, there can be but two possible solutions to the problem. There are two different mean distances of least possible error, — one of which is 36, and the other 30. The one is included within the theory and limits of Le Verrier, and corresponds with Adams' solution; the other is the orbit of Neptune. The simple view of the case . . . reconciles all the computations and observations, as well as the discords and contentions. It does not detract in the slightest degree from the well-earned fame of the illustrious geometers who had arrived at a solution of the problem, and I am not aware that it has ever been opposed by mathematical reasoning.

Another critic of the Adams–Le Verrier solution was the French astronomer E. M. Liais. He actually made his comments in 1866, and they were published in 1892. The quotes which follow are from my translation from the French. (Liais 1866).

Liais maintained that the credit for Neptune's discovery should go neither to Le Verrier nor Adams, but to Alexis Bouvard — even though Bouvard did not live to see the identification of the planet (he died in 1843). In 1834, wrote Liais, Bouvard discussed the problem with his nephew Eugène. Eugène Bouvard's tables of Neptune, presented to the Institute though not actually published until 1845, were much better than those of Alexis. Eugène felt that there were indications of a distant planet which had been in conjunction with Uranus in 1822. His conclusions were:

(1) There must be a planet perturbing Uranus, and it must be remote, as otherwise it would cause measurable effects upon the movement of Saturn.
(2) The perturbations were not detectable all through Uranus' revolution; they were inappreciable between 1690 and 1800.
(3) The perturbations were greatest when Uranus and the new planet were in conjunction. This dates the conjunction as having occurred in 1822.
(4) The action of the new planet was detectable for about 25 years before and 25 years after conjunction with Uranus. Therefore, it should be possible to find the distance of the planet, and hence its longitude, for any given time.

As there were no measurable perturbations between 1690 and 1800, continued Liais, it was in 1690 that an interval of 25 years had passed since the last conjunction with Uranus, and this gives the date of the conjunction as 1665. The interval between 1665 and 1822 is 157 years, which must be the interval between successive conjunctions. As the period of Uranus is 84 years, an interval of 157 years between successive conjunctions gives a revolution period for the unknown planet of 180 years, corresponding to a distance of 32 astronomical units; the lower limit for the distance must be set at 28 astronomical units, as otherwise the perturbations would be detectable for longer. From all this, the longitude of the planet on 1 January 1847 could be set between 323 degrees and 333 degrees. Actually, it was 329 degrees.

In his calculations, commented Liais, Adams followed Bode's Law, and used methods of approximations, while Le Verrier tried to determine the elements of the planet solely by correlations with the elements of Uranus. Le Verrier concluded that the distance of the planet could not be less than 35 astronomical units. In this he was wrong; and though his longitude was near the truth,

this fortuitous result was due to the fact that his researches were made at a time not far from the last conjunction, so that Neptune had not travelled very far since ... If the last conjunction had taken place much earlier, and Le Verrier had used the same erroneous value for the distance, Neptune would have turned up a long way from its calculated position. Therefore, it is with reason that we must discuss the agreement between Le Verrier's position and the real one as being due to chance ... Le Verrier frequented the Paris Observatory every day — so why did he not make use of the telescopes there? He did not dare to suggest it. None of the Paris astronomers had enough confidence in his work after reading it, particularly as the author himself was not attempting a search. No doubt for the same reason, Arago took no action. ... Therefore, all the circumstances tend to confirm the work of Bouvard and contradict that of Le Verrier. In fact, Le Verrier's hypothetical planet, near the position where Neptune was found, does not exist, and must be relegated to the realm of fiction. ... We must conclude that it is to Alexis Bouvard that we must give the honour of the geometrical discovery of Neptune, and to Galle the optical discovery of this hitherto unknown celestial body.

Liais also adds that Le Verrier's work 'gives an example of the remarkable series of errors which can arise by plunging blindly into analysis before taking all the circumstances into account'.

Liais' paper seems to be somewhat biased, to put it mildly. It is surely wrong to claim that he did not dare to recommend a search from Paris; there is every reason to suppose that he was anxious for this to be done, and only after the French inertia did he write to Galle at Berlin. Moreover, Alexis Bouvard's tables were far from flawless, and contained errors detected by both Le Verrier and by Airy. Also, the perturbations of longitude were dependent chiefly upon the mass of Neptune, the distance between it and Uranus, and the differences in the heliocentric longitudes. So despite their errors with regard to the distance from the Sun, both Adams and Le Verrier were on the right track. If, as Liais and Peirce claimed, the agreement had been sheer luck, it would have been stretching coincidence rather too far.

Of course, similar types of arguments have been used much more recently with regard to the discovery of Pluto in 1930, and more will be said about this later. But one fact cannot be denied. The first men to discover Neptune and recognize it for what it was — a planet, not a star — were neither Bouvard, nor Adams, nor Le Verrier. They were Johann Gottfried Galle and Heinrich D'Arrest.

Over the next few years the orbital elements of Neptune were refined and improved, until well before the end of the nineteenth century they had been defined very accurately. Meanwhile, attention was being paid to the physical aspect of the planet, and to discuss this we must begin with the curious episode of William Lassell and the ghostly Neptunian ring.

4

Lassell's ring

William Lassell was one of the most distinguished amateur astronomers of the last century — and even though we must discount the sprained ankle story, he was very much concerned with the early studies of the planet. He did, indeed, discover the larger satellite, Triton, a few weeks after Neptune itself had been identified. But Lassell also believed that he had detected a ring, and in this, as we now know, he was mistaken.

Lassell was born in Bolton in 1799. He became a brewer, and in this career he was successful, so that he was able to devote all his leisure time to astronomy. He made his own mirrors — the first, a 7-inch (17.8-cm), in 1820 — and by 1844 he was able to begin constructing a 24-inch (61-cm), which, for that period, was very large indeed. He had the help of James Nasmyth, an astronomer who was also the inventor of the steam-hammer and who later co-authored, with James Carpenter, the first major British book dealing with the surface of the Moon. Lassell also sought advice from the Earl of Rosse, who was puting the finishing touches to his great 72-inch (183-cm) reflector at Birr Castle in Ireland, and who certainly knew more about mirror-making than anyone else at the time. Lassell's telescope was a success. In 1852 he transferred it to Malta, where the observatory conditions were much better than those at his home near Liverpool, and while there he made important observations of many kinds.

As we have seen, the news of Galle's discovery of Neptune reached London on 30 September 1846. On 1 October, John Herschel wrote to Lassell asking him to look for any satellites 'with all possible expedition', using the new 61-cm reflector. Lassell lost no time. He discovered Triton on 10 October, but his first observations had been made as early as the 2nd, as soon as he received Herschel's request. (Letters were delivered more quickly in those days than they are now!) It was on the 2nd that he saw not only Neptune, but also suspected that it had a ring. Further observations made during the following weeks made him confident that the ring really existed (Lassell 1846a).

Much earlier, after the discovery of Uranus by William Herschel, a Uranian ring had been reported by the discoverer himself. Nobody else ever saw it, and there is no doubt that it was an illusion, due to problems with the optical arrangement of William Herschel's telescope. But in the case of the alleged Neptunian ring, there was apparent confirmation, which is why the whole matter is worth examining more closely. Two outstanding accounts of it have appeared in recent times: one is by R. M. Baum (1973) and the other by R. W. Smith and R. M. Baum (1984). In preparing the following chapter I have naturally consulted original sources, but in fact Baum and Smith have carried out the research so exhaustively that there is little left to be done, and, inevitably, I have drawn upon them heavily in the account which follows. All the Lassell references not precisely defined are in the Lassell papers in the archives of the Royal Astronomical Society, and are listed by Smith and Baum (pp. 15–17).

In Lassell's diary for 1846 we read: 'I observed the planet last night, the 2nd, and suspected a ring . . . but could not verify it. I showed the planet to all my family and certainly tonight have the impression of a ring.' On the 3rd he 'received the impression of a ring, not much open, and nearly at right angles to the parallel of daily motion' (Lassell 1846a). On 10 November: 'The planet very like Saturn, as seen with a small telescope and low power, but much fainter.' On these nights several people observing with him saw the supposed ring, and all in the same direction, as shown by independent diagrams. He had 'never looked at the planet, under tolerable circumstances, without receiving the same impression' of the ring's existence, and so far as he could judge, 'the direction of the ring made a constant angle with the meridian, not with the horizon, though this was not very certain'.

Lassell had recently described his observations in a letter to the *Times*, dated 12 October and published on the 14th.

> With regard to the existence of the ring, I am not able absolutely to declare it, but I received so many impressions of it, always in the same form and direction, and with all the different magnifying powers, that I feel a very strong persuasion that nothing but a finer state of atmosphere is necessary to enable me to verify the discovery. (Lassell 1846b)

In the following month he presented a report to the Royal Astronomical Society (Lassell 1846c). In view of his excellent reputation, it was natural for other astronomers to accept his findings, particularly inasmuch as he was using a very large telescope. One of those present at the Royal Astronomical Society meeting at which Lassell spoke was John Russell Hind, who had been making regular observations of Neptune with the 7-inch Dollond refractor at George Bishop's observatory in Regent's Park. On 11 December 1846 Hind made his first report of the ring. 'The existence of a ring appears as yet undecided, though most probable. [The planet] presents an oblong appearance in Mr. Bishop's refractor' (Hind 1847).

James Challis then entered the lists — and, one has to admit in retrospect, with no more success than before. He made his first observation of the ring on 12 January 1847, with the Northumberland refractor. He wrote:

> I had for the first time a distinct impression that the planet was surrounded by a ring. The appearance noticed was such as would be presented by a ring like that of Saturn, situated with its plane very oblique to the direction of vision. I felt

convinced that the observed elongation could not be attributed to atmospheric refraction, or to any irregular action on the pencils of light, because when the object was seen most steadily I distinctly perceived a symmetrical form . . . . I saw the ring again on 14 January. In my notebook I remark: 'The ring is very apparent with a power of 215, in a field considerably illumined by lamp-light. Its brightness seems equal to that of the planet itself.' . . . The ratio of the diameter of the ring to that of the planet, as measured from the drawings, is about that of 3 to 2. The angle made by the axis of the ring with a parallel of declination, in the south-preceding or north-following quarter, I estimated at 60 degrees . . . . I am unable to account entirely for my not having noticed the ring at an earlier period of the observations. It may, however, be said that an appearance like this, which it is difficult to recognize except in a good state of the atmosphere, might for a long time escape detection, if not expressly and repeatedly looked for. (Challis 1847a)

Challis informed Lassell, who was naturally elated, and replied: 'I cannot refuse to consider that your observation puts beyond reasonable doubt the reality of mine; especially as even your measured angle of position agrees with my estimation within 4 degrees' (Lassell 1847). Challis was by now fully convinced of the existence of the ring, though he did again stress that he was 'unable to account for my not having noticed the Ring earlier' (Challis 1847b).

Very vague reports of something unusual in the aspect of Neptune came also from W. C. Bond and M. F. Maury in America and from F. di Vico in Italy, but all these are so indefinite that they cannot be regarded as even partial confirmation of the ring. Bond, using the 15-inch (38-cm) refractor at Harvard, the first reasonably large telescope in the United States, wrote: 'There is no appearance of a ring to Neptune when viewed with high powers, though with low powers there seems to be an elongation' (Bond 1848). If the 'elongation' had been genuine, Bond would certainly have noticed it more clearly with a higher magnification. Therefore, the presence of a ring depended mainly upon the observations by Lassell, Hind and Challis.

Doubts soon began to creep in. In particular, the Rev. W. R. Dawes, nicknamed 'the eagle-eyed' because of his keen vision, was dubious, as he said in a letter written to Challis as early as 7 April 1847. On the other hand, the Council of the Royal Astronomical Society declared that 'the existence of the ring seems almost certain' (Royal Astronomical Society 1847) and then, in September, Dawes visited Lassell at Starfield, Lassell's observatory near Liverpool, and for the first time claimed that when using the 24-inch telescope he had glimpsed the ring on several occasions.

In 1848 Lassell made fresh observations, and during August, together with Janes Nasmyth, he 'received as strong an impression of the ring as at any previous view'; Nasmyth agreed. But later, in September, he was unable to see it, and he was no more successful during 1849, though admittedly the general weather conditions were poor. By 1850 it seems that he was coming to the conclusion that the ring did not exist, but he could not be sure, and he hoped to find out for certain when he took his telescope to the clearer skies of Malta.

On 4 November 1852, from Malta, he recorded that he had 'an impression of an extremely flattened ring . . . . I think I have never seen Neptune so well before', though he added: 'I suspect some illusion' (Lassell 1852a). Finally, on 15 December, he found that the supposed ring appeared to change in position when he rotated the tube of the telescope.

> It is thus evident that the phenomenon keeps a constant angle with the direction of the telescope and not at all with the parallel of declination, proving that whatever may be the cause it is more intimately related to the telescope than the object. (Lassell 1852b)

This was the report. Lassell never again suggested that Neptune had a ring in any way like that of Saturn, and there were no further observations from Hind, Challis, Dawes or anyone else. The ring of Neptune, like Herschel's ring of Uranus, passed into the realm of scientific myth. But why did it seem, initially, to be so convincing?

The first report was Lassell's, but eventually Lassell himself came to the reluctant conclusion that faults in his telescope were to blame for the illusion. He had found that there were problems in preventing the large primary mirror from flexing under its own weight as the telescope was moved around in altitude and in azimuth, and this was a defect which he was never able to eliminate completely. Flexure of the primary would be precisely the sort of fault which would distort the shape of a small object such as the disk of Neptune, producing an elongated or even ring-like appearance. Lassell may have realized it. Note, too, that as soon as he became convinced that the ring was non-existent, he did not hesitate to say so and to admit that he had been misled.

Apart from the observations made with Lassell's reflector, the only confirmations came from Challis and Hind. Neither, in all probability, considered the idea of a ring until after Lassell's announcement of it, and so they went to their telescopes with preconceived ideas. It is painfully easy to 'see' what one expects to see — remember the canals of Mars! — and it is difficult to avoid the conclusion that they were guilty of what has often been described as wishful thinking. The observations by Dawes and Nasmyth were made with Lassell's 24-inch, and it is surely significant that the keen-sighted Dawes was never able to glimpse the ring with any other telescope.

After the lapse of so many years, it is safe to say that whatever Lassell saw, it was not a true ring, and is not in any way connected with the ring system discovered in modern times. Nobody would accuse the observers of dishonesty, any more than one would accuse Percival Lowell of dishonesty in covering Mars with a network of canals. It is only too easy to make a mistake, and the inital belief in the ring of Neptune was one of Lassell's comparatively rare errors of judgement.

Thanks to the Voyager 2 mission, we now know that Neptune does indeed have a ring system (see Chapter 13), but it has no connection with Lassell's non-existent ring. It is virtually unobservable from Earth, and certainly no telescope available in Lassell's time could possibly have shown even a trace of it.

Plate 1 – Portrait of the German mathematician Frederick Wilhelm Bessel, of Königsberg. From the Frontispiece of 'Complete Works', (R. Engelmann, Editor), published Leipzig in 1875. Photograph reproduced courtesy Royal Astronomical Society Library.

Plate 2 – Portrait of Sir George Biddell Airy, Astronomer Royal and Director of the Royal Greenwich Observatory. From R.A.S. MS 94 No. 58. Photograph reproduced courtesy Royal Astronomical Society Library.

Plate 3 – Portrait of the brilliant mathematician, John Couch Adams (1819–1892) who, after graduating in 1843, remained at the University of Cambridge and turned his attention to tracking down the planet which was acting upon and disturbing Uranus. By mid-1845, Adams had obtained an approximate position for the new planet (Neptune). From an engraving by Cousins of a portrait by Mogford. Photograph reproduced courtesy Institute of Astronomy, Cambridge.

Plate 4 – Portrait of James Challis (1803–1882), Plumian Professor of Astronomy at the University of Cambridge and Director of the Cambridge Observatory between 1836 and 1861. Photograph reproduced courtesy Institute of Astronomy, Cambridge.

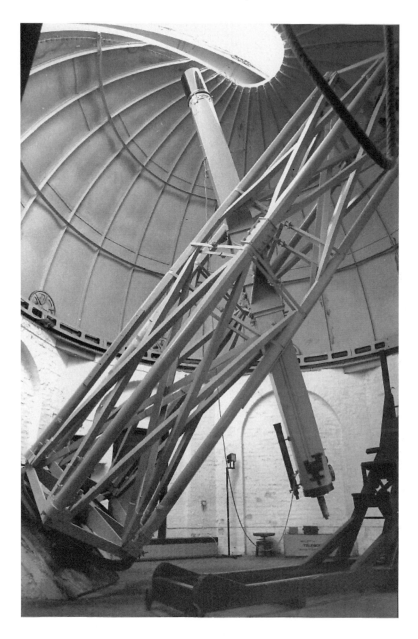

Plate 5 – The Northumberland refractor, with its 29.8-cm object-glass, at the Cambridge Observatory, used by Challis in the search for Neptune, and viewed here from the east. Photograph reproduced courtesy Institute of Astronomy, Cambridge.

Plate 6 – Portrait of the French chemist turned mathematician, Urbain Jean Joseph Le Verrier who, although adopting a different approach to John Couch Adams in his calculations of the position of the planet which was acting upon and disturbing Uranus, arrived at a final result which was almost the same.

Plate 7 – Portrait of Johann Gottfried Galle, of the Berlin Observatory, to whom Le Verrier wrote in 1846 asking him to begin searching in the position indicated for the new planet (Neptune). From R.A.S. MS 91 Vol 3 No. 6. Photograph reproduced courtesy Royal Astronomical Society Library.

Plate 8 – Portrait of Johann Encke, Director of the Berlin Observatory, where Galle and D'Arrest carried out their successful search for Neptune in September 1846.

Plate 9 – Portrait of Heinrich Louis D'Arrest, an enthusiastic young student astronomer, who asked to be allowed to join Galle in the search for Neptune at the Berlin Observatory.

Plate 10 – The 23-cm Fraunhofer refractor used by Galle and D'Arrest to identify Neptune on 23 September 1846. It was then at the Berlin Observatory, but is now in the museum at Munich. Photograph courtesy of Patrick Moore.

Plate 11 – Portrait of William Lassell, a brewer by profession, who became an expert observer and built a private observatory near Liverpool. He discovered Triton, Neptune's largest satellite, soon after the planet itself had been identified. From R.A.S. MS 91 Vol 3 No. 9. Photograph reproduced courtesy Royal Astronomical Society Library.

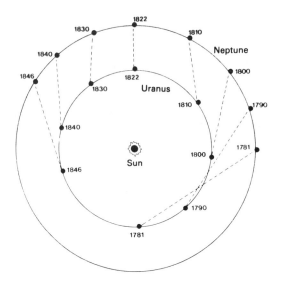

Plate 12 – The effects of Neptune on Uranus between 1781 (the year of Uranus' discovery) and 1846. Neptune was in opposition with respect to Uranus in 1822; before that date Uranus was accelerated by Neptune's gravitational pull, and subsequently it was retarded.

Plate 13 – Photograph of Neptune and its satellites, taken using the 60-inch reflector at Mount Wilson. Triton (arrowed) appears just below the bright image of Neptune, while Neried (also arrowed) appears towards the upper right of the picture. Photograph reproduced courtesy of the Mount Wilson Observatory.

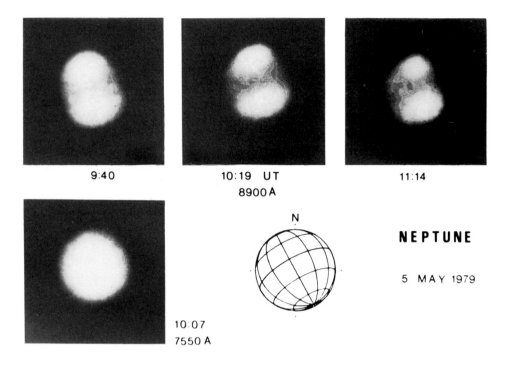

9:40   10:19 UT   11:14
8900 Å

N

NEPTUNE

5 MAY 1979

10:07
7550 Å

Plate 14 – A series of four photographs of Neptune, taken using the 154-cm reflector at the Catalina Observatory of the University of Arizona, on 5 May 1979, by Brad Smith, Harold Reitsema and Stephen Larson, showing cloud features on the planet. Photograph reproduced courtesy of University of Arizona.

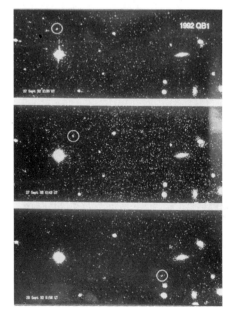

Plate 15 – Sequence of three images taken on 27–29 September 1992, showing the motion of the first 'Kuiper Belt' object, designated 1992 QBI (encircled dot), obtained by David Jewitt and Jane Luu, who discovered this asteroidal sized trans-Neptunian object using the 2.2 metre reflector on Mauna Kea, Hawaii. Its distance from the Sun varies between about 5086 million and 6582 million kilometres, and its orbital period is 296 years.

Plate 16 – Photograph of Clyde Tombaugh, who discovered the ninth planet, Pluto, in 1930, after a painstaking and systematic search carried out at the Lowell Observatory in Flagstaff, Arizona. Photograph courtesy of Patrick Moore.

5

Neptune as a planet

We come now to events which span the period between the discovery of Neptune, in 1846, and the Voyager 2 pass in 1989. In this chapter I propose to deal in turn with the size and shape of the planet, its rotation period, and its brightness and variability.

1. Diameter, form and mass

Even though Neptune is a giant planet, it is so far away that its apparent diameter is not at all easy to measure accurately, and a small error in measurement can cause a dramatic change in the linear value adopted. Obviously the first efforts were made by using conventional micrometers, but it is surprisingly difficult to define the precise edges of a small disk whose outlines are not particularly sharp. All that could really be said of the earlier determinations was that they were of the right order — between 2 and 3 seconds of arc — corresponding to a real diameter of around 50 000 kilometers. Until very recent times it was never certain whether Neptune was slightly larger than Uranus, or slightly smaller. We now know that it is smaller, but decidedly more massive. (For the record, the mean apparent diameter of Uranus is 3.7 seconds of arc, giving the real diameter as 51 118 kilometres.)

G. E. Taylor (1969) has given a list of some determinations of the apparent diameter of Neptune, ranging from the mid-nineteenth century through to modern times; to these I have added (see Table 3) some of the earlier determinations, some of which were (understandably) considerably too great.

In the same paper — Taylor's Presidential Address to the British Astronomical Association — he described another method of measuring the apparent diameter of a small disk: by using occultations. Clearly, the duration of the occultation of a star, which is essentially a point source, will give a value for the diameter of the occulting body, provided that the distance and velocity are known. Taylor himself has pioneered this method, from 1952, and has found it to give excellent results, notably with asteroids. The drawback, of course, is that one has to wait for Nature to provide a suitable occultation. By now there have been numerous attempts with asteroids and with planetary satellites. So far as Neptune is concerned, there have been measurements using the occultation method, notably by Kovalevsky and Link, Bixby and van Flandern, French *et al.,* and Hatanaka.

Table 3

Observer(s)	Method	Diameter (seconds of arc)
Encke (Berlin, 1846)	Micrometer	3.2
Laugier (1846)	Micrometer	2.60
Main (1846)	Micrometer	4.36
Challis (Cambridge. 1847)	Micrometer	2.99
Hind (London, 1847)	Micrometer	2.47
Mädler (Dorpat, 1847)	Micrometer	2.40
Struve (Pulkovo)	Micrometer	2.19
Lassell (1853)	Micrometer	2.71
Lassell and Martin (Malta, 1867)	Double-image micrometer	2.24
Kaiser (Leiden, 1872)	Double-image micrometer	2.87
Seeliger (Munich)	Micrometer	2.53
Barnard (Lick)	Micrometer	2.43
See (Washington)	Micrometer	2.01
Wirtz (Strasbourg)	Micrometer	2.31
Rabe (Wroclaw)	Micrometer	2.25
Camichel (Plc du Midi)	Diskometer	2.09
Kuiper (McDonald)	Diskometer	2.05
Kuiper (McDonald)	Double-image micrometer	2.02
Dollfus (Pic du Midi)	Double-image micrometer	2.04

It seems that all the results obtained by the various methods are reasonably consistent. At the General Assembly of the International Astronomical Union, held at Baltimore in August 1988, M. E. Davies and his collaborators presented the results of the official Working Group concerning the dimensions and rotational properties of planets and satellites (Davies *et al.* 1988). They gave a value for the equatorial diameter of Neptune as being 50 538 km, as against 51 118 km for Uranus, with a possible uncertainty of no more than 20 km. A selection of past and more recent determinations is given in Table 4.

It is interesting to note that some of these measurements which would be expected to be the most reliable — for example, by Kuiper and Camichel — are actually the worst. They also over-estimated the mean density of the globe; Kuiper gave a specific gravity of 2.22 and Camichel 2.17, as against Barnard's older value of 1.33 and Taylor's recent determination of 1.54. Incidentally, W. De Marcus (1977) stated that the possible error in the diameter measurements was of the order of 5 per cent, but it turned out to be rather more than that.

The flattening of the globe is measurable, though in 1894 Barnard could not detect it with the Lick refractor even though he found the flattening of Uranus perceptible. G. E. Taylor, in 1969, gave the equatorial diameter as 50 940 km, but adopted 49 920 km for the polar diameter. These values are much the same as those

Table 4

Observer(s)	Date	Diameter (equatorial) (km)
Dick	1852	49 880
Denison	1866	61 570
Chambers	1889	59 860
Barnard	1902	52 900
Moore and Menzel	1928	50 000
Kuiper	1949	44 600
Camichel	1953	45 000
Taylor	1968	49 800
Taylor	1969	50 940
Kovalevsky and Link	1969	50 450
Dollfus	1970	48 600
Taylor	1970	49 800
Bonev	1972	45 810
I.A.U.	1977	47 600
French et al.	1985	50 536
Taylor	1986	49 500
B.A.A. Handbook	1988	48 400
Davies et al.	1988	50 538

given by the 1988 Working Group (Davies 1988). As we have seen, their derived value for the equatorial diameter was 50 538 km; the polar value was 49 600 km, with a possible uncertainty of 30 km. The corresponding values for Uranus were 51 118 km and 49 946 km. This means that the equatorial diameter of Neptune is actually greater than the polar diameter of Uranus, and the two globes really are almost alike, though of course Neptune is appreciably the more massive of the two.

The current value for the oblateness of Neptune is 0.02, as given by Sharonov (1964), Taylor (1969) and French (1984). Kovalevsky and Link (1969) gave 0.021, with a possible uncertainty of 0.004, as a result of the occultation by Neptune of the star BD−17°4388. The axial inclination is 28°48′.

Incidentally, A. H. Cook (1979) reconsidered some recent data on the geometrical and dynamical flattening of both Uranus and Neptune in the light of new determinations of the rotation periods, and found some of them to be physically implausible.

Neptune's mass is now thought to be 1/19 314 that of the Sun, or 17.2 times that of the Earth (appreciably more than that of Uranus, which is 1/22 869 that of the Sun or 14.6 times that of the Earth). The surface gravity is 1.22 times that of the Earth. Some older mass determinations were as shown in Table 5. Most of these values, apart (understandably) from the first, are reasonably close to the truth. Though Neptune does not equal Uranus in actual size, it is quite definitely the third most massive member of the planetary system.

Table 5

Year	Authority	Method	Mass (Sun=1)
1847	O. Struve	Motion of Triton	1/43 866
1848a	B. Peirce	Motion of Uranus	1/20 000
1848b	B. Peirce	Motion of Triton	1/18 780
1849	J. C. Adams	Motion of Triton	1/17 900
1850	A. Struve	Motion of Triton	1/14 491
1851	G. P. Bond	Motion of Triton	1/19 400
1855	J. R. Hind	Motion of Triton	1/17 135
1862	J. Safford	Motion of Uranus	1/20 039
1874	S. Newcomb	Motion of Uranus	1/19 700
1875	S. Newcomb	Motion of Triton	1/19 560
1876	E. Holden and A. Hall	Motion of Triton	1/18 520
1891	C. Young	Various	1/19 380

2. Rotation

All the giant planets have relatively short rotation periods — less than 10 hours in the case of Jupiter, less than 11 hours for Saturn. Measuring the periods of the various features on Jupiter is easy enough, because one only has to watch them. Saturn is less forthcoming, but for many years the rotation period has been known reasonably well. With both these planets, the equatorial rotation period is appreciably shorter than the period in higher latitudes.

Uranus and Neptune present much greater problems, because of the lack of surface features — coupled, in the case of Uranus, with the extraordinary axial tilt. Until recently, therefore, all that could be done was to look for slight, regular variations in magnitude which could be attributed to rotational effects, or else to use the spectroscope and measure the Doppler shifts at alternative limbs together with the inclinations of the spectral lines.

Early estimates were much too short, and the first measurement which could be treated with any confidence at all was that of J. H. Moore and D. H. Menzel, who used the spectroscopic method (Moore and Menzel, 1928). They gave a value of 15.8 hours, with a possible uncertainty of one hour either way. Since the true value now seems to be 16.1 hours, this was a very good effort in view of the extreme difficulty of the observations.

Some values which have been given up to the present time are listed in Table 6.

Surprisingly, the value given in the *Handbook of the British Astronomical Association* for 1995 was still '18 to 20 hours', which certainly seems at variance with the best data available. (The value for Uranus was given as '16 to 28 hours'.)

E. Öpik and R. Liviander made their estimate by using light variations, as reported by Maxwell Hall, who had obtained a value of 7 hours 55 minutes many years earlier (Hall 1884a). Öpik and Liviander found a rotation period for the equatorial zone of 7 h 42 m 42 s.1, increasing to 7 h 50 m 10 s.7 for the temperate

Table 6

Date	Authority(ies)	Period (hours)
1872	Flammarion	10.97
1884	Hall	7.914
1924	Öpik and Liviander	7.7
1928	Moore and Menzel	15.8±1.0
1930	Moore and Menzel	15.8±1.0
1977	Hayes and Belton	22±4
1978	Cruikshank	18.17 or 19.58
1978	Slavsky and Smith	18.44±0.01
1980	Münch and Hippelein	11.8±1.2
1980	Belton, Wallace, Hayes and Price	15.4±3
1980	Brown, Cruikshank and Tokunaga	17.95
1983	Terrile and Smith	17.83±5
1984	Belton and Terrile	18.2±0.4

zones, giving a double periodicity combined with a total absence of polar flattening. They concluded that Neptune was the quickest spinner of the giant planets (Öpik and Liviander 1924). In fact, all these conclusions were wrong, and it was surely straining credibility to work out rotation periods to within a fraction of a second when the actual error amounted to well over 100 per cent!

S. H. Hayes and M. J. S. Belton (1977a, b) made observations of the tilts of the spectral lines, and found the rotation periods to be 24±3 hours for Uranus and 22±4 hours for Neptune. Neptune, unlike Uranus, was found to spin in the direct sense, in agreement with almost all earlier results. There were, however, misgivings in the case of Neptune because a relatively long rotation period did not seem to fit in with the measured value for the oblateness. Subsequently Belton, Wallace, Hayes and Price (1980) confirmed that there had indeed been errors of interpretation, due mainly to possible limb darkening, and that the real period was likely to be 15.4 hours, though with an uncertainty of 3 hours. It was assumed that the period must lie somewhere between 15 and 20 hours, and it was added that 'all observers agree that the rotation period of Uranus is longer than that of Neptune'.

Using light-curves, D. P. Cruikshank (1978) derived two plausible values for the rotation period — either 0.7572 day or 0.8160 day, each with an uncertainty of 0.002 day. In the same year D. Slavsky and H. J. Smith recorded that 'on the basis of the apparently valid assumption that quasi-stable longitudinal photometric inhomogeneities exist in the atmosphere of Neptune, the rotation period associated with the principle inhomogeneity is found to be 18.44±0.01 hours' (Slavsky and Smith 1978), in satisfactory agreement with Cruikshank. E. M. Drobyshevskii supported the idea of equatorial currents with quicker rotation; he considered that these could explain the discrepancies in the published values, assuming that the jets had widths up to 80 degrees and a rate 25 to 40 per cent faster than that of the polar regions (Drobyshevskii 1979). Later, Belton and Terrile found what they termed a clear signature of large-scale, zonal atmospheric flows in the Neptunian atmosphere, with wind velocities at least as great as 109 metres per second; there was no evidence, pro or con, for atmospheric motions on Uranus (Belton and Terrile 1984).

A new method of determination was introduced by B. J. Terrile and B. A. Smith, in 1984. For the first time, a period was derived from observations of cloud features high in Neptune's atmosphere. The actual observations were carried out in May 1983, with the 2.5-metre Irénée du Pont reflector at the Las Campanas Observatory in Chile, together with a CCD camera and a broad-band infrared filter centred at 8900 Ångströms; at this wavelength the images appear dark where methane deep in Neptune's atmosphere absorbs sunlight. Obvious features were made out, sometimes down to a resolution of half a second of arc. Their analysis yielded a period of 17 hours 50 minutes, with a possible uncertainty of 5 minutes (Terrile and Smith 1984). This was only 7 minutes shorter than the value found photometrically a few years earlier by Brown, Cruikshank and Tokunaga (1980). Before the Voyager pass of Neptune, the value found by Terrile and Smith was regarded as the best available. *En passant,* the rotation period of Uranus is now known to be 17 hours 14 minutes.

3. Brightness and variability

The mean opposition magnitude of Neptune is 7.8, so that it is never visible with the naked eye, though binoculars will show it easily. Quite apart from its changing distance from us, there could be variations due to changes in the Neptunian atmosphere, or to changes in the amount of light sent to it from the Sun.

As early as 1884, an attempt to work out a rotation period from changes in magnitude was made by Maxwell Hall, of the Kempshot Observatory in Jamaica. Between 29 November and 14 December 1883 he found that the planet was a delicate bluish hue, changing in magnitude between 7.6 and 8.3. He wrote:

> Respecting the cause of these variations, the planet may be assumed to have dark belts on its surface similar to the belts on Jupiter and Saturn; should these belts break up so as to make one hemisphere darker than the other, the effect of axial rotation would be to produce variations in brightness above and below the mean, as is observed, and then the period of variation would simply be the period of axial rotation. (Hall 1884a)

His derived rotation periods ranged between 8.2 and 9.2 hours, later amended to 7.914 hours (Hall 1884b). (As we have noted, it was this alleged variation which later led Öpik and Liviander to derive a completely erroneous rotation period.)

Hall's results were not confirmed elsewhere. At Harvard, E. C. Pickering, using a photometer and taking the star λ Ursæ Minoris as a comparison, found that the magnitude was always between 7.5 and 7.7 (Pickering 1884, 1885a). At Potsdam, G. Muller obtained a similar result (Muller 1884). A detailed criticism of Hall's results was made by J. M. Baldwin, who used a Zöllner photometer together with the 13.5-cm Steinheil refractor; between 25 January and 28 April 1908 he found that the magnitude was always from 7.9 to 8.0 (Muller 1884). He concluded that

> the variations of the estimated brightness of Neptune announced by Hall are probably due to errors of observation, as the measurements of other observers

at about the same time, and also at other oppositions, give not the slightest trace of any such variation. Consequently the time of rotation of 7.9 hours deduced by him must be regarded as unsupported by observation.'

On the system adopted in the Potsdam Dürchmusterung, the mean opposition magnitude of Neptune was 7.99; Muller gave 7.97, Pickering 8.00.

There have, however, been more recent claims that Neptune's brightness — and also the brightness of Titan and Uranus — is affected by the changing output of the Sun. In 1977, G. W. Lockwood found that between 1972 and 1976 all three bodies brightened by from 0.005 to 0.025 magnitude per year; no definite explanation was offered, but 'a possible influence of solar activity upon planetary albedo is suggested by the fact that all of the objects observed have brightened during the declining half of the solar cycle' (Lockwood 1977). Further comments along the same lines were made in the following year (Lockwood 1978). During the following cycle, Lockwood and D. T. Thompson (1986a) found that 'the visual brightness and albedo of Neptune varies periodically during the 11-year solar cycle with an amplitude of 4 per cent, anti-correlated with the variation of solar ultraviolet output. A seasonal trend in colour suggests that Neptune, like Uranus, may have a slightly reddened pole.'

Atmospheric changes, leading to variability, have been investigated recently by D. P. Cruikshank, who finds that the upper atmosphere of the planet is variable with at least four characteristic time-scales. Global inhomogeneities in high-altitude haze distribution remain stable for several days, and allow measurements of the rotation period to be made, but this stability sometimes breaks down, completely obscuring the diurnal light-curve. He also confirms the apparent long-term, low-amplitude variability of magnitude in anti-correlation with the solar cycle, as found by Lockwood and his collaborators, and he feels that this low-amplitude variability is occasionally punctuated by outbursts which have their greatest contrast in the near infrared (Cruikshank 1984, 1985). The high-altitude condensation of particles in the Neptunian atmosphere may have a decay time of several months.

The fluctuations of Neptune seem, therefore, to be definite; they are very slight, and cannot be detected without delicate instruments, but they are worth following, and they may be able to provide us with a great deal of interesting information.

The Voyager 2 results have shown that the mean density of Neptune's globe is 1.77 times that of water, so the estimates of Kuiper and Camichel were rather high, and Taylor's determination was low but not too wide of the mark. The rotation period is 16.1 hours so that most of the values given in Table 6 are too long; Moore and Menzel were the most accurate. Note, however, that Neptune has differential rotation (see Chapter 13).

6

Structure and appearance of Neptune

Of the four giant planets, Jupiter and Saturn make up one pair while Uranus and Neptune make up another. Jupiter differs from Saturn, and Uranus differs from Neptune, but the differences between the Jupiter–Saturn pair and the Uranus–Neptune pair are much more significant.

Jupiter and Saturn consist mainly of hydrogen and helium; they are thought to have silicate cores, surrounded by layers of liquid hydrogen either metallic (at great depths) or molecular (higher up), with the extensive 'atmospheres' above. Uranus and Neptune have less hydrogen and helium, but much more water, ammonia (NH_3) and methane (CH_4). It is now regarded as certain that the planets were formed from the original solar nebula — the old, attractive idea that they were pulled off the Sun by a passing star has been rejected — and we have a good idea of the age of the Solar System, because the age of the Earth is known to be of the order of 4.6 thousand million years. Of course, the planets took a long time to evolve. J. A. Fernandez and W. H. Ip have calculated that Neptune took about 100 million years to acquire its present mass (Fernandez and Ip 1981).

M. Podolak and A. G. W. Cameron (1974) have given estimates for the composition by 'weight' of the Sun and the giant planets: these are shown in Table 7.

Table 7

	Density (Water=1)	Weight per cent		
		Hydrogen (H_2) and helium	Water, ammonia and methane	Iron, silicates and oxides
Sun	1.7	98.4	1.2	0.3
Jupiter	1.3	82	5	13
Saturn	0.7	67	12	21
Uranus	1.2	15	60	25
Neptune	1.7	10	70	20

Slight revisions to the density values of the planets do not make any major difference to the general conclusions. The percentage of the two lightest gases, hydrogen and helium, decreases with increasing distance from the Sun, while the percentage of 'ices' of H_2O, NH_3 and CH_4 increases. It may be said that the Uranus–Neptune pair forms a sort of link between the very large, hydrogen-rich Jupiter–Saturn pair, and the rocky inner planets, though it would be dangerous to take this statement too literally.

Undoubtedly the root cause of the difference lies in the greater distance of Uranus and Neptune from the embryo Sun. They never accreted cores as large as those of Jupiter or Saturn, and were therefore unable to pull in as much of the hydrogen-rich gas of the solar nebula. A suggestion by V. M. Slipher that there might be a constituent gas lighter than hydrogen never met with much support (Slipher 1904).

Because Neptune is so remote, and shows no obvious surface detail, little was known about it until modern times. Until well into the twentieth century, it was still thought likely that all the giant planets, including Neptune, were miniature suns, giving off enough heat to warm their satellite systems. This idea was disproved by Sir Harold Jeffreys in 1923, who showed that a more likely picture was that of a cold gaseous layer above a solid surface, with the clouds having a lower melting point and a lower boiling point than water (Jeffreys 1923).

Rather different models were proposed in 1947 by Rupert Wildt, who assumed that each giant planet had a central rocky core, overlaid by a thick layer of ice above which came the gaseous atmosphere (Wildt 1947). For Neptune, the core would have a diameter of 19 300 km, with the ice-layer 9600 km thick and the atmosphere extending for 3200 km. Shortly afterwards, W. R. Ramsey proposed models which were different again. According to this picture, Jupiter and Saturn were made up mainly of hydrogen, which in the lower layers was metallic. The lesser masses of Uranus and Neptune had resulted in the loss of much of their original hydrogen and helium, so that they were composed largely of water, methane and ammonia (Ramsey 1951). So far as the two outer planets are concerned, it seems that Ramsey's theory is not very wide of the mark, but it has been refined and modified in recent years. It should not be surprising for the bulk of their material to be made up of 'ices' (i.e. substances which would be frozen if they lay near the planet's surface); these ices would be expected to be of water, ammonia and methane, which, because they are compounds of the most reactive elements (hydrogen, oxygen, carbon and nitrogen) are the commonest ices in the Solar System.

A detailed theory of the interiors of Uranus and Neptune was described in 1980 by W. B. Hubbard and J. J. MacFarlane. This was a three-layer model, with each planet having a central rocky core, a mantle made up of water, methane and ammonia 'ices', and an outer envelope made up chiefly of hydrogen and helium. The model incorporated a new H_2O equation of state, based on exponential data, which was considerably 'softer' than previous H_2O equations of state; the interior temperatures were assumed to be of the order of 5000°C (Hubbard and MacFarlane 1980). M. Podolak and R. T. Reynolds presented a series of models in which the relative amounts of rock, ices, and hydrogen/helium were varied; by taking the density and gravitational force into account, the compositions of the globes could be worked out, though several possible models were considered because of uncertainties as to the rotation periods (Podolak and Reynolds 1981). Later, the same authors, together with R. Young, considered the difference between the internal structures of Neptune and Uranus; if they really were as similar in structure as they are in mass and diameter, it followed that the rotation period of Neptune would have to be shorter than that of Uranus (Podolak, Young and Reynolds 1985).

An interesting idea was proposed in 1981 by Marvin Ross, who suggested that Uranus and Neptune might contain layers of diamond. At a temperature of from 2000 to 4000°C, and a pressure of a million atmospheres, methane will dissociate into hydrogen and carbon, and the carbon could become compressed into crystals of diamond. If the carbon did not dissolve into its surroundings, crystals of diamond would precipitate toward the planet's centre, possibly encasing the core with a layer thousands of kilometres thick (Ross 1981). This is certainly a novel suggestion!

According to the current picture, there is a rocky core 16000 km across, surrounded by a mantle of 'ices' in liquid form made up of water, methane and ammonia, over which comes the low-density hydrogen–helium atmosphere. The rocky core is between 4 and 12 times as massive as the Earth (Hubbard and MacFarlane 1980) and the pressure is perhaps 20 million atmospheres, with a temperature of 7000°C. But there is one difference between Uranus and Neptune which is extremely important. Neptune, like Jupiter and Saturn, has a strong source of internal heat; Uranus apparently has not.

If there were no major heat-source from below the surface, Neptune would have a temperature of around 44 K — that is to say, −229°C. Actually it is 57 K or −216°C, and this is virtually the same as that of Uranus, which is over 1600 million kilometres closer to the Sun and is of comparable mass. No doubt Uranus' unusual axial tilt has also something to do with this discrepancy (Hunten 1984), but it seems that it is Uranus, not Neptune, which is the 'odd one out', since both Jupiter and Saturn have very pronounced internal heat-sources.

The cause of Neptune's heat-source has often been discussed. L. Trafton has shown that it cannot be due to frictional dissipation of the tides raised in Neptune by the large satellite Triton (Trafton 1974), and therefore processes inherent in the globe are presumably responsible. The liquid or partly-liquid mantle round the core, made up of 'ices', is in convective motion, carrying up heat from the core region. Apparently Neptune gives off 2.4 times as much energy as it would do if there were no internal heat-source (Murphy and Trafton 1974, Stier *et al.* 1977, Neff 1985 and Hubbard 1978). There have been suggestions that the cause of the heat-sources in giant planets is slow contraction, plus the unmixing of the icy constituents in the mantle so that the heavier materials sink downward toward the core, but the whole question remains decidedly open — and Jupiter, with its exceptionally large mass, is probably a special case. But that Uranus and Neptune differ markedly in this respect is not in doubt.

Not much can be learned by ordinary telescopic observations of Neptune. The disk is detectable with small apertures, as was pointed out long ago by M. A. W. Holmes, using a 6-inch (15-cm) reflector (Holmes 1908), but generally the disk appears featureless; even Edward Emerson Barnard commented that he could see no surface markings with the great refractors of Lick and Yerkes (Barnard 1894). T. J. J. See claimed to have seen equatorial belts, using the 26-inch (66-cm) refractor at Washington (Clerke 1902), but whether or not this can be accepted is dubious; See, a strange and highly unpopular character, also reported seeing craters on Mercury (Gordon 1983). I know of only one drawing of Neptune which shows anything definite; this was by Thomas Cragg, using a magnification of ×1000 on the Mount Wilson 60-inch (152-cm) reflector on 17 April 1953. Cragg showed a bright equatorial band, with duskier poles (Cragg 1953). G. E. Taylor (1969) commented that

it is difficult to observe any details even with the largest telescopes. Although a few rather ill-defined markings have been observed, they have been inadequate for the determination of a period of rotation. Limb darkening is quite noticeable. .... The slightly greenish colour we observe is due to the absorption of the longer wavelengths by a very extensive atmosphere.

B. A. Smith and H. J Reitsema justifiably conclude that 'the disks of Uranus and Neptune are probably featureless in visible light' (Smith and Reitsema 1982).

I have looked at Neptune through large telescopes on various occasions, and I cannot claim to have seen any surface features, even though on one occasion — on 3 May 1983, using the 24-inch (61-cm) reflector at the Mount John Observatory at Lake Tekapo, New Zealand — seeing-conditions were well-nigh perfect, at least in my experience. While Uranus is definitely green, I always see Neptune as bluish. I suppose it all depends upon the nature of one's eyes!

Using filters and CCD equipment, B. A. Smith, H. J. Reitsema and S. M. Larson have recorded cloud features (Smith, Reitsema and Larson (1979), but the best results of the pre-Voyager period were obtained by B. A. Smith and R. Terrile with the 2.5-metre Irénée du Pont reflector at Las Campanas in Chile, referred to in the previous chapter. Using a CCD camera and two coronagraphs, direct images of both Uranus and Neptune were taken through a 8900 Å (890 nm) filter, 40 nm wide, centred on the methane absorption band. The image reveals a cloud pattern similar to that seen in 1979.

The 1983 Neptune images are characterized by several patches of haze located in the mid-latitudes in both northern and southern hemispheres. Because the Earth's declination at Neptune is presently at −24 degrees, the southern-hemisphere features are most clearly seen. Neptune appears to show less limb brightening in the 890 nm methane band than did Uranus during the mid- and late 1970s, but this may be due in part to the preliminary processing; Neptune appears to have cloud structure with inherently higher contrast, and the images, therefore, are not as strongly enhanced during processing. (Smith 1984)

From the same set of images, it is concluded that the rotation is prograde. And also in 1984, J. H. Heasley and his collaborators found that CCD images at 8900 Å showed that conditions on Neptune are decidedly variable, a fact which they attributed possibly to the planet's rotation (Heasley, Pilcher, Howell and Caldwell 1984).

Finally, and very much of an aside, what are the prospects of any life in or on Neptune?

From Neptune, the light of the Sun would be about the same as that of 687 full moons as seen from Earth — that is to say, almost 80 times the light of a standard candle seen from a distance of one metre, or an electric arc viewed from a few metres (Young 1891). Neptunian astronomers would have a poor view of the other planets; the maximum elongation of Venus from the Sun would be a mere 1½ degrees, Mars 3 degrees and Jupiter 10 degrees. Saturn would be a naked-eye object, though it must be remembered that Neptune is much further away from Saturn than we are; only Uranus and Pluto would be better seen than they are from Earth. Even Uranus would be out of view for long periods, during the time when it and Neptune were on opposite sides of the Sun, and the same would apply to Pluto. G. F. Chambers, whose delightful last-century *Handbook of Astronomy* is still of interest and value today, wrote that

Though deprived of a view of the principal members of the Solar System, the Neptunian astronomers are well circumstanced for inspecting comets, and are also able to take, probably with considerable success, observations on stellar parallax, seeing that they are in possession of a base-line 584 700 000 000 miles in length, or 30 times the length to which we are confined. (Chambers 1861, p. 83).

In the fourth edition of his *Handbook*, he also has a column detailing 'Time required for falling into the Sun, were gravitation free to exert its influence alone'. The Earth would be consumed in a mere 64.6 days, but for Neptune the delay time would be 10 640.2 days, or over 29 years! (Chambers 1890.)

More seriously in the light of present-day knowledge, L. Margulis and his collaborators have examined the possibilities of the growth of micro-organisms on Uranus, Neptune and Titan. After examining a wide range of physical, chemical and other conditions, they conclude that the growth likelihood of contaminant terrestrial micro-organisms on Uranus is nil, so that 'if, as is likely, conditions are even more extreme on Neptune, the probability of contamination of both the outer planets is nil' (Margulis, Halvorson, Lewis and Cameron 1977). This, at least, is a conclusion not likely to be challenged in the future.

It may be worth adding that in 1992 I was able to observe Neptune with the Palomar 152-cm reflector, under good conditions, and was quite unable to see any trace of surface features.

7

The atmosphere of Neptune

Like the other planets, Neptune has long been known to have a massive atmosphere above its mantle and core, as described in a useful review by R. G. Prinn (1973). In 1969 J. Kovalevsky and K. Link estimated that the atmosphere extended downward to within 24 200 km of the centre of the globe, though these figures require some revision today (Kovalevsky and Link 1969). However, the lack of similarity between Uranus and Neptune is also evident with their atmospheres, as has been pointed out by D. M. Hunten. 'There are differences in heat flow, stratospheric thermal structure and composition, and spectral appearance. Presumably these are traceable to some combination of Uranus' unusual obliquity and similar distance from the Sun, but most of the details are obscure' (Hunten 1984). The atmosphere of Uranus is clear to great depths, while, as H. P. Larsen has stressed, that of Neptune has a cloud layer of variable height (Larson 1979). There is a haze of ice crystals or aerosol particles, contributing to a warming of the upper atmosphere by the absorption of sunlight and producing a temperature inversion of the same type as is found on Saturn's satellite Titan, which, as was established by the Voyager 1 mission, has a dense atmosphere composed chiefly of nitrogen. At times, nearly half of Neptune is covered with haze, which dissipates and re-forms in a matter of days or weeks. As we have seen, there is also an anti-correlation with the solar cycle; at solar maximum, Neptune's brightness fades, so that events on the Sun may affect the formation and dissipation of haze in Neptune's atmosphere.

The clouds visible at some wavelengths have already been described.

On Neptune there are enough distant clouds to show up in images and to permit estimations of the rotation period. Such work is done at wavelengths strongly absorbed by methane, to suppress light scattered by the enormous amount of hydrogen in the atmosphere. It is assumed that the clouds in question must be at high altitudes, perhaps even stratospheric, to be made visible in this way. (Hunten 1984)

And from researches carried out by M. G. Price and O. G. Franz, observations in the 7300-Ångström methane band require the presence of an optically thin layer of brightly scattering aerosol particles high in the Neptunian atmosphere (Price and Franz 1980).

Of course the main constituent of the atmosphere is hydrogen, as was established by Herzberg long ago (G. Herzberg 1951), but the amount of methane in relation to hydrogen is several times greater than in the Sun (W. Macy and L. Trafton 1975a). The thermal inversion due to the absorption of sunlight by methane is an extra proof of Neptune's internal heat-source, since otherwise methane would have to be supersaturated by an unrealistically large factor. The stratospheric temperature inversion seems to be stronger with Neptune than with Uranus (H. Moseley, B. Conrath and R. F. Silverberg 1985). It is also worth noting that methane and ethane emissions are prominent in the spectrum of Neptune, but not in that of Uranus (W. Macy and W. Sinton 1977a); the presence of methane was established more than half a century ago, by Rupert Wildt (Wildt 1937).

The methane clouds are of particular interest. It seems that stratospheric clouds of methane and other hydrocarbons are being transported convectively through the cold trap, which would be consistent with the variability of the clouds. Quite possibly, convection in some parts of the Neptunian atmosphere is so strong that it 'overshoots' the regular radiative–convective boundary, and delivers methane gas into the stratosphere on a time-scale shorter than the time-scale for precipitation. There are, however, other possibilities. Very small particles of methane ice could be updraughted into the stratosphere, where they re-sublimate; the atmosphere could be so stable that methane would remain in a supercooled state in the lower temperature region, passing from the troposphere to the stratosphere in a gaseous state. In some regions the temperature minimum might not be low enough to produce a cold trap at all, so that, in effect, the cold trap would have 'leaks' (G. S. Orton and J. F. Appleby 1984, p. 136). The first of these explanations seems to be the most likely. Strong updraughts could carry the precipitated methane crystals straight through the temperature-minimum region into warmer altitudes, where they can re-evaporate (Hunten 1984, p. 31). On Uranus, on the other hand, the cold trap seems to be effective.

Yet though Neptune has a rather unexpectedly warm thermal inversion layer, recent work due to A. T. Tokanuga, G. S. Orton and J. Caldwell seems to indicate that it may be rather less pronounced than has been thought (Tokunaga, Orton and Caldwell 1983). Further investigations are needed.

Again and again we come back to the all-important fact that Neptune has a strong internal heat-source, while Uranus has not; the effective temperatures of the two are much the same, and are of the order of 58 K ($-215°C$). Thus M. T. Stier and his colleagues, using results from a 102-cm telescope carried aloft in a balloon, give 59.7 K for Neptune and 58.5 K for Uranus (Stier, Traub, Fazio, Wright and Low 1978); G. S. Orton and his collaborators give 58.1 K for Neptune and 54.1 K for Uranus (Orton, Baines, Bergstralh, Brown, Caldwell and Tokunaga 1987), while H. Moseley, B. Conrath and R. F. Silverberg (1985) give 58.1 K for Neptune and 57.7 K for Uranus. The temperature on Neptune seems to be fairly uniform, with little difference either between different latitudes or for different seasons (L. Wallace 1984). J. Veverka and his colleagues state that the diurnal temperature variation is less than 10 degrees, and may be zero; the average temperature is less than 15 degrees lower at latitude 55 degrees than it is at Neptune's equator (Veverka, Wasserman and Sagan 1974).

M. J. S. Belton, L. Wallace and S. Howard have shown that there is definite evidence of 'weather' on Neptune, with systems of zonal winds (Belton, Wallace and Howard 1981). Observations by R. R. Joyce and his collaborators, made between April 1975 and March 1976, indicated that over a period of one year, an extensive high-altitude cloud layer formed over the planet and then partially dissipated (Joyce, Pilcher, Cruikshank and Morrison 1977) — this was in fact the first really good evidence for atmospheric activity on Neptune. More recently, it has been suggested by B. N. Khare *et al.* that some types of complex dark hydrocarbon products are formed in Neptune's clouds by cosmic rays, and that there is indirect evidence of this from 'a darker, slightly red circumpolar region on Neptune' (Khare, Sagan, Thompson, Arakawa, and Votaw 1986). The shapes of the methane bands in Neptune's spectrum are unlike those of Uranus; instead of having flat, almost black bottoms, they are almost pointed in the middle, and the best match is with Saturn's major satellite Titan. 'The resemblance of Neptune's and Titan's spectra suggests that Neptune's atmosphere may contain an extended, somewhat absorbing haze' (Hunten 1984, p. 35).

Finally, what about the full extent of the atmosphere? Measurements were made in 1986, when Neptune passed in front of the star BD − 17°4388 and occulted it. It was found that photoelectric measures gave a value for the occultation equatorial diameter of 49 800 km, corresponding to the upper reaches of the atmosphere of 50 830 km with an uncertainty of 140 km; this was increased to 50 940 km, with the same uncertainty, if a correction for the deflection of light in gravitational field was applied.

All in all, the atmosphere of Neptune is of exceptional interest. It is different from any other in the Solar System, and is even different from that of its non-identical 'twin', Uranus.

Voyager 2 showed that the winds on Neptune are indeed very strong, and a great deal of new information about the atmosphere was obtained (see Chapter 13), modifying some of the conclusions given here.

8

Radio waves and the Neptunian magnetosphere

More than thirty years ago, Jupiter was found to be a source of radio emission; later it was established that there is a close association with the highly volcanic satellite Io. The Jovian magnetosphere is very extensive — at times is can even engulf Saturn! — and of course the magnetic field is very powerful; the planet is also associated with radiation zones which would be lethal to any astronaut. (Io, which lies within these zones, is a world which must always be viewed from a respectful distance.) Saturn, too, is a radio source, though less powerful than Jupiter. The Saturnian magnetosphere extends out to the orbit of its satellite Titan, which is sometimes within the magnetosphere and sometimes outside it.

Obviously, radio emissions from Uranus and Neptune are much more difficult to detect. With Neptune, the first positive results were obtained in 1966 by K. Kellerman and I. I. K. Pauliny-Toth, who observed at a wavelength of 1.9 cm (Kellerman and Pauliny-Toth 1966). A few years later, observations at 11.1 and 3.7 cm indicated that the microwave temperature might be rather higher than expected (W. J. Webster *et al.* 1972). Observations made with the 85-ft NRL radio telescope at Maryland indicated that there was increasing atmospheric transmission from 3 mm to 3 cm, plus a temperature increase with depth in the atmosphere to almost 200 K ($-73°C$) (C. H. Mayer and T. P. McCullough 1971).

Before the Voyager 2 mission, it was well-nigh impossible to prove the existence or non-existence of magnetospheres associated with either Uranus or Neptune, but various theories were proposed. Despite Neptune's strong internal heat-source, it was suggested that there might be no magnetic field at all, because of the high density and pressure at the core (R. Smoluchowski and M. Torbett 1981). Yet suppose that the field originated not inside the core, but outside? There was the possibility that the pressure variation and the associated conductivity of water was not likely to be enough to satisfy the conditions necessary for a dynamo on Uranus, but was marginal for Neptune, so that the expected presence of metallic water in a thick layer around Neptune's core might make the operation of a dynamo more plausible there. Some sort of convective activity was likely, and this might be expected to occur largely in the icy mantle, compressing the water and ionizing it. If so, it might conduct electricity well enough to produce a dynamo in the mantle, making Neptune unique in having a magnetic field generated not inside its core, but well outside. There was also Triton to be taken into account — a large, massive satellite, moving in a retrograde sense and already thought to have an appreciable atmosphere. If Io were associated with Jupiter's magnetosphere, why should Triton not be linked with Neptune's, particularly if there were any chance of weak icy volcanic activity there? (G. L. Siscoe 1979).

Shortly before the Voyager 2 pass of Uranus, a detailed review of the possible Uranian and Neptunian magnetospheres was given by T. W. Hill. He concluded that only 'a rare combination of circumstances' could conspire to invalidate the conclusion that 'externally-driven convection should be unimportant in the internal dynamics of the magnetospheres of Uranus and Neptune', unless Triton really did play a major rôle in the case of Neptune. He also commented that 'from all the above arguments, one might well conclude that the magnetospheres of Uranus and Neptune, if indeed they exist, are rather boring places' (Hill 1984, pp. 512 and 514).

Then, in January 1986, came the Voyager encounter with Uranus, and the positive detection of both radio emission and a magnetosphere (see Hunt and Moore 1988). The magnetic field was detected five days before Voyager's closest approach, at a distance of 275 Uranian radii from the planet — that is to say, almost 7 million kilometres. It proved to be anything but 'boring'. At a distance of 450 000 kilometres Voyager entered the magnetosphere, revealing a strange magnetic dipole field with an axis tilted by almost 60 degrees to the rotational axis of the planet — and, moreover, offset from the centre of the globe. The Uranian magnetosphere is now known to extend to 500 000 kilometres on the day side of the planet, and 6 million kilometres on the night side, so that all the satellites are included in it. (Oberon, the most distant, moves at 573 500 kilometres from Uranus.)

Possibly the fact that Uranus' magnetic axis is so different from the rotational axis is linked with the unique axial inclination — a characteristic which Neptune, of course, does not share. However, the fact that Uranus did indeed turn out to have a powerful magnetic field was an indication that Neptune probably had one too. The first definite indications were obtained by Imke de Pater and Michael Richmond of the University of California, Berkeley, when they used the Very Large Array radio telescope in 1982 and 1986 to search for synchrotron radiation. They detected the telltale signal above Neptune's radio noise by comparing their observations with what had been predicted by atmospheric models. Neptune's magnetic field was found to be ten times weaker than Jupiter's, but probably two to three times stronger than that of Uranus (de Pater and Richmond, 1988). However, final proof had to await the arrival of Voyager 2 in August 1989.

Voyager 2 showed that Neptune's magnetic field is weaker than those of the other gas giant planets (see Chapter 13). Surprisingly, the magnetic axis is inclined to the rotational axis by 47 degrees, so that in this respect Neptune resembles Uranus.

9

Rings or arcs?

Saturn has always been called 'the ringed planet', and certainly its ring-system is not only superbly beautiful but is also unique. The Voyager missions have shown it to be far more complex than had been expected, with thousands of ringlets and narrow divisions. The particles making it up are icy, and it has a very high albedo.

Not until 1977 was it found that Uranus, too, has rings, and we know that Jupiter also has a system of sorts, though it is too obscure to be seen from Earth, and is quite different in nature from the rings of Saturn. But for the space missions, Jupiter's thin, dark rings would certainly have remained undetected. Not so with Uranus, where the rings were discovered by skilled observation — admittedly more or less by accident.

Mention has already been made of the occultation technique pioneered by Gordon Taylor, in which the diameter of a small disk can be measured by finding out how long it takes to pass over a background star. Taylor actually began his programme in 1952, but it was not until 22 January 1977 that he had his first real triumph. Following his calculations, observations of the occultation of the 8.8-magnitude star SAO 156867 by Uranus were made from several sites, including a team on board the Kuiper Airborne Observatory. Both before and after occultation by the planet, the star showed symmetrical 'winks' which could hardly be due to anything but a system of Uranian rings. Confirmation was soon forthcoming from other occultations and, of course, from the Voyager 2 pass of January 1986. Ten thin, dark rings and a broader sheet of material are now known, with distances from Uranus ranging between 37 000 km and 51 160 km. The two inner satellites, Cordelia and Ophelia, act as 'shepherds' to the outer ring — known as the ε-ring, which has a width of from 22 to 93 kilometres and a markedly eccentric form.

Using infrared techniques with the 3.9-metre Anglo-Australian Telescope at Siding Spring, D. A. Allen was the first to produce useful pictures of the rings as seen from Earth (Allen 1983). Operating at wavelengths of from 2 to 2.4 microns, he produced a striking view of the ring system (Allen 1984). Previously, the tremendous light-grasp of the Hale reflector at Palomar had been needed to show anything at all. What, then, were the prospects for Neptune?

Visually, the outlook was not particularly promising. Allen's attempts to detect indications of Neptunian rings were not successful. As he wrote,

> The results were disappointing. The planet is too small to see clearly, and shows no hint of a ring . . . . And to complicate the issue, the K window (2.0 to 2.4 microns) does not always find Neptune dimmed by methane absorption, for on some occasions the planet shows a high-altitude haze which reflects sunlight quite well and brightens the planet several-fold. The haze is, I suppose, a bit like cirrus clouds in our atmosphere. (Allen 1984)

It seemed, then, that the only hope of detecting Neptunian rings, pending the arival of a space-craft, was to use Taylor's occultation method.

Of course, the situation was obviously different from that with Uranus. Quite apart from the fact that Neptune does not share Uranus' extraordinary axial tilt, there is the presence of a close-in large satellite, Triton, which has retrograde motion (that is to say, moving in a sense opposite to that of the rotation of the primary). All the other known retrograde satellites are small and presumably asteroidal, so that Triton is an exceptional case. There were grounds for suggesting that its presence might make conditions unstable enough to prevent the formation of a ring. A. R. Dobrovolskis investigated the situation mathematically, and found that stable rings could exist at inclinations of 0 to 15 degrees, 165 to 180 degrees, or around 90 degrees to Neptune's equator, but that perturbations by Triton would produce severe warping of the ring-plane. He concluded that if Neptune did possess rings, they might not lie in the plane of the equator (Dobrovolskis 1980).

In fact, the first significant observation — if it can be regarded as significant! — was obtained as long ago as 1968 by several observers, notably a team from Villanova University. They were studying the occultation of a 7th-magnitude star by Neptune, not in a quest for rings (nothing of the sort had ever been contemplated at that time) but to measure the apparent diameter of the disk. E. Guinan and J. S. Shaw observed from Mount John Observatory in the South Island of New Zealand, and other records came from Japan and Australia. Nothing of particular note was found, and the observations were more or less forgotten — until 10 May 1981, when two occultations were studied by a whole team of investigators in a definite search for rings.

It then occurred to the Villanova team to take another look at their 1968 data. Difficulties were encountered in locating the Mount John records, but eventually the tracing of the event was reconstructed, and it was found that there was a well-defined fade of the star, lasting for $2\frac{1}{2}$ minutes, which began about three minutes after emersion from behind Neptune's disk. The extent of the fade amounted to 0.3 of a magnitude, and the dip corresponded to a distance, on the equatorial plane, of from 28 550 to 32 560 kilometers from the planet's centre (Guinan, Harris and Maloney 1968).

It was interesting, but not conclusive, and Guinan was suitably cautious, even suggesting that there might be unstable rings with lifetimes of only a few hundreds of thousands of years — not long on the cosmical scale. At any rate, the technique was clearly worth following up, and was tackled energetically by, among others, J. L. Elliot, who had been a key member of the team identifying the rings of Uranus in 1977. On 24 May 1981 observers using the telescopes at Catalina and Mount Lemmon, in Arizona, observed the close approach of Neptune to a star, and found that there was a very brief drop in the star's brightness — obviously not due to a ring, but, they suggested, possibly to an unknown satellite with a diameter of about 180 kilometres (Elliot and Kerr 1984, p. 185; H. L. Reitsema *et al.*, 1981).

At the spring meeting of the American Astronomical Society in 1982, Guinan and his colleagues reported their findings with the old 1968 occultation. An over-enthusiastic journalist was moved to write, in the *New York Times*, an article headed 'Data show Two Rings Circling Neptune: Astronomer says they appear to be 1200 miles wide — Make-up Undetermined' (Wilford 1982). This was certainly prema-ture, but future events were awaited with keen interest.

Another occultation took place on 15 June 1983, and was carefully observed by several teams. W. B. Hubbard and his colleagues obtained data from six sites in the south-west Pacific, and searched for rings down to a distance of 0.03 Neptune radii from the planet's surface, but without success (Hubbard *et al.* 1985). Elliot's team also obtained negative results, and observations from Mauna Kea in Hawaii and from Mount Stromlo and Siding Spring in Australia, as well as from the Kuiper Airborne Observatory, showed

no evidence for equatorial rings beteen 25 300 and 200 000 kilometres .... These results rule out a Neptunian system similar to that of Saturn or Uranus, but not a system of low optical depth similar to the Jovian rings. The data show no features that appear likely to have been caused by material in the equatorial plane of Neptune near the Roche limit. (Elliot *et al.* 1985)

At an occultation on 12 September 1983, an Indian team working at the Uttar Pradesh State Observatory on Manora Peak, Naini Tal, recorded four brief fadings at 1.16, 1.20, 1.22 and 1.28 Neptune radii from the centre of the planet (A. K. Pandey, H. S. Mahra and W. Mohan 1984), but these observations were not confirmed elsewhere, and it is fair to say that they were generally treated with some reserve, even though the authors believed that they indicated the presence of a ring system extending from 64 190 to 64 400 kilometres in Neptune's equatorial plane (Pandey and Mahra 1987).

We come next to the occultation of the star SAO 186001, on 22 July 1984. This time the results were more positive. At the European Southern Observatory in Chile, F. Gutirrez, J. Manfroid, R. Häfner and R. Vega observed a 'wink' as the star faded by 65 per cent; it was suggested that the occulting body was a satellite 10 to 15 kilometres in diameter, at 75 000 kilometres from Neptune. P. Bouchet and his colleagues used the 1-metre and 0.5-metre telescopes, and found that the dimming was seen in both, ruling out the possibility of instrumental error (Manfroid, Häfner and Bouchet 1986). A hundred kilometres to the south, at Cerro Tololo, Hubbard and his team made photoelectric measurements with the 0.5-metre reflector, and found a 32 per cent drop in brightness lasting for about 1.2 seconds (Hubbard *et al.* 1986). This was attributed to a partially transparent arc of material at a distance of 67 000 kilometres from Neptune — the first suggestion of an 'arc' rather than a complete ring. A detailed description of the events was given by R. Kerr.

Hubbard observed the bluish planet and the reddish star at three different wavelengths. Only something near Neptune would block the light from the star and not the planet. The event happened about 0.1 seconds later in the south than in the north [bearing in mind that Cerro Tololo is south of the European Southern Observatory], as would be expected for a true ring occultation. The star lost 35 per cent of its light, corresponding to a ring 10 to 15 km wide and 70 000 to 80 000 km from Neptune's centre. But why was no dimming seen on the outer passage? Does this indicate an arc? (Kerr 1984)

Reitsema made the appropriate comment that he was 'willing to accept part of a ring, but that creates real problems, because I don't understand how you get parts of rings'.

The idea of arcs seemed to take hold, and more searches were made during the occultations of 7 and 25 June 1985. The star occulted on the first occasion was a binary, and since both components were occulted it was possible to find their relative positions very accurately. A single sharp dip was observed after emersion, and if this were due to material near Neptune it would be either 62 600 km or 63 760 km from the planet (assuming it to lie in the equatorial plane), depending upon which of the binary components was occulted. The results seemed to tie in well with Hubbard's of 1984, but only one dip was noted, so that presumably the other component of the binary escaped. Neither were there any phenomena before immersion. The 25 June occultation was also successfully observed; this time, no evidence for ring-like material was found in the region examined during post-emersion, which included the entire range of equatorial radii over which events had previously been reported (Covault, Glass, French and Elliot 1986).

As Voyager 2 continued on its long journey toward Neptune, various papers about the alleged 'arcs' appeared. It was claimed, for instance, that 'recent analyses of isolated events observed around Neptune indicate the existence of an incomplete or at least a highly azimuthally variable ring or arc round Neptune' (B. Sicardy et al. 1986). It was also suggested that an undiscovered satellite might play an important rôle. P. Goldreich and his colleagues proposed that the incomplete rings consist of a number of short arcs centred on the co-rotation resonances of a single satellite, which would have a radius of around 100 kilometres and would move in an inclined orbit (Goldreich, Tremaine and Borderies 1986). And J. J. Lissauer referred to hypothetical 'shepherd satellites', and suggested that the incomplete ring-arc could be azimuthally confined near a Trojan point of an unknown moon. Two satellites of 100 to 200 kilometres in diameter would be sufficient to confine the ring, and of course these would be too small and faint to be detectable from Earth (Lissauer 1985). There is nothing revolutionary in this; after all, some rings of Saturn and Uranus are known to have 'shepherds'.

On the other hand, it had to be admitted that the very existence of ring-arcs was unproved. Even if they existed, it was likely that the presence of Triton would make the situation very different from anything found in the systems of the other giant planets. Voyager 2 would, it was hoped, clear up the problem of whether Neptune had a complete ring, a series of arcs, or no ring-system at all. As two of the chief investigators, Elliot and Kerr, commented in 1984, 'If Neptune has rings, they will almost certainly not be discovered from the ground.'

Complete rings do indeed exist (see Chapter 13), but the outermost ring includes three arcs of denser material, where there is more dust than in the rest of the ring. One of the pre-Voyager occultation events, that of 24 May 1981, seems to have been due to a chance observation of the then-unknown small satellite, Larissa.

10

Triton and Nereid

Though William Lassell almost certainly did not know about the hunt for Neptune until it was too late for him to join in, he was not tardy in starting observations of the new planet. In 1846 he discovered the main satellite (Lassell 1846) and made frequent observations of it during the following months mainly in his attempts to verify the existence of a ring. He wrote:

> Unfavourable weather, and the low altitude of the planet, have not allowed Mr. Lassell to observe the ring of Neptune satisfactorily: there is, however, no doubt in his mind as to the existence of a ring. The observations of the satellite have been more successful: it has been seen repeatedly in the course of the year, and the non-existence of any star in the places successively occupied by it frequently ascertained. From the mean of his observations, Mr. Lassell concludes that the satellite revolves about the planet in 5 hours 21 minutes nearly, and that its greatest elongation is about 18 seconds of arc. The orbit which it appears to describe has a minor axis differing little from the diameter of the planet.
>
> The satellite is much brighter in the preceding than in the following half of its path. The sixth satellite of Saturn varies similarly in brightness. This periodical variation seems to shew that one side of the satellite has less power of giving back light than the other, and that the time of rotation upon its axis is equal to its periodic time round the planet, as is the case in our own moon. (Lassell 1847).

The actual revolution period is 5 d 21 h 3 m, and the elongation is indeed about 18 seconds of arc, so that in this respect Lassell's description was accurate enough. As we have noted, the motion is retrograde. At mean opposition distance the magnitude is about 13.6 — almost half a magnitude brighter than Titania, the brightest of the satellites of Uranus.

Lassell's statement that the satellite is much brighter in one half of its orbit than in the other has not been borne out. This is certainly true of Saturn's satellite Iapetus (referred to by Lassell as the sixth satellite, now usually regarded as the eighth), which has large areas covered with dark deposit, for reasons which are still obscure. Iapetus is two magnitudes brighter when west of Saturn than when to the east. This is not true of Triton, but it is quite correct to infer, as Lassell did, that the revolution period is the same as the axial rotation period, so that Triton keeps the same face turned permanently toward Neptune. The orbit is practically circular.

The name Triton, incidentally, was mentioned long ago (Fouche 1905) and is said to have been originally proposed by the French astronomer Camille Flammarion; in 1932 H. N. Russell said that it 'might well achieve general recognition' (Russell 1932). For many years the name was regarded as unofficial, in the same way as those of the four Galilean satellites of Jupiter, but it is now always used, and is certainly appropriate mythologically.

Because Triton is relatively bright, even at its immense distance, it must be either highly reflective, extremely large by satellite standards, or both. The first attempt at accurate measurement was made in 1954 by G. P. Kuiper, using the Hale reflector at Palomar; he gave a diameter of 3800 km, making Triton slightly larger than our Moon but appreciably smaller than Titan in Saturn's system or Ganymede and Callisto in Jupiter's. Subsequent estimates varied considerably. Some were as high as 6000 km, others as low as 2500 km; the later value was obtained by speckle interferometry at the Canada–France–Hawaii telescope on Mauna Kea (D. Bonneau and F. Foy 1986). Before the Voyager pass, the generally-adopted value was of the order of 3500 km (D. P. Cruikshank 1984, S. F. Dermott 1984), though L. Trafton made it 4200 km (Trafton 1984a). The IAU Working Group of 1988, chaired by M. E. Davies (Davies 1988), gave a value of 3500 km, with a possible uncertainty of 500 km. Triton was then thought to be the largest known planetary satellite apart from Ganymede and Callisto in Jupiter's system and Titan in Saturn's. It is certainly considerably larger than Pluto, a fact which may well be significant in considering the status of Pluto itself.

Trafton gave the mass of Triton as 1.4×10^{26} g. This is about 1/800 that of Neptune. By contrast, the reciprocal mass values for the most substantial satellites of Jupiter, Saturn and Uranus, taking the primary as unity, are 1/12 820 (Ganymede), 1/4150 (Titan) and 1/20 000 (Titania). Apart from our Moon, and excluding the unique Pluto–Charon pair, this would have made Triton the most massive satellite in the Solar System with respect to its primary. (*En passant*, our Moon, with a mass 1/81 that of the Earth and a diameter more than a quarter of that of the Earth, way well be regarded as a companion planet rather than as a mere satellite.) Triton was also thought to be exceptionally dense; R. Greenberg gave it a specific gravity of over 4 (Greenberg 1984).

It is Triton's retrograde motion which makes it unique among large satellites. As we have noted, the other known retrograde satellites are probably asteroidal, but Triton is much too large to be a captured asteroid. Obviously we are dealing with a very remarkable object indeed.

Whether Triton must be classed as a formerly independent body which was 'captured' by Neptune, whether its orbit was violently disturbed in the early history of the Solar System, or whether it is simply a normal satellite with an abnormal motion is not yet clear. Deferring discussion of this for the moment (see Chapter 11) there have been suggestions that the current orbit is contracting, so that Triton has a limited life-span. The first hint of this was given almost a century ago (Struve 1894). It has been subsequently claimed that Triton will reach the Roche limit in from 10 to 100 million years, and will break up (T. McCord 1966). In 1984 this was questioned by A. W. Harris on the grounds that Neptune might itself have retrograde rotation, in which case the distance of Triton would grow instead of decrease (Harris 1984b), though we are now certain that Neptune rotates in the direct sense, so that Harris' argument does not apply.

Until fairly recently, the only planetary satellite known to have an atmosphere was Titan; this was detected as long ago as 1944, by Kuiper, though its constitution was not known until the Voyager 1 pass; Titan's atmosphere is mainly nitrogen, with a ground pressure considerably greater than that of the Earth's air at sea-level. The Galilean satellites of Jupiter have no atmospheres of this sort. Triton, however, is both massive and cold, so that the existence of an atmosphere seemed plausible, even though an early search by H. Spinrad failed (Spinrad 1969).

By 1988 the presence of an atmosphere had been firmly established. It was generally believed to be composed chiefly of methane, with a good deal of nitrogen, and this affected our ideas about the surface conditions. D. P. Cruikshank wrote that 'near-infrared spectrophotometry of Triton and Pluto at low spectral resolution and signal precision reveal methane absorption on both bodies. The absorption on Triton is probably gaseous CH_4, while that on Pluto is a combination of gas and ice of CH_4' (Cruikshank 1984). Earlier, it had been suggested that the surface was likely to be rocky (or dusty) rather then frosty:

> We picture a surface that is largely covered with rocky material with a few patches of frozen CH_4, probably away from the sub-solar point, where the temperature is lower. The dark side of Triton may act as a cold trap; because of the inclination of the satellite's orbital plane to that of Neptune, portions of the body are in perpetual darkness during part of Neptune's orbit round the Sun. The variability of the geometry over the 165-year period, together with the satellite's apparent 5.877-day rotation period, may result in a methane meteorology, the details of which merit further study. (Cruikshank and Silvaggio 1979)

The computed surface partial pressure of CH_4 (was 1 ± 0.5) $\times 10^{-4}$ bar, a value consistent with the calculated vapour pressure of methane gas above methane ice at the expected surface temperature of Triton. It was, however, added that there was 'no compelling evidence' for the presence of solid methane on the surface.

Certainly there must be large seasonal variations in the mass and composition of the atmosphere, as was pointed out by L. Trafton.

> Condensed phases of gases making up the bulk of Triton's atmosphere are likely to exist on Triton's surface in the form of solid or liquid 'polar caps' which extend as far as 55 degrees from the poles. The mass of Triton's atmosphere is governed by the energy balance between the sunlight these caps absorb and the heat they radiate to space. The polar cap temperatures should be approximately equal and uniform over their surfaces. Because of the rapid precession of Triton's orbit around Neptune's pole, the insolation and, therefore, the temperature of the polar caps must vary in a complex fashion. This will cause the mass of Triton's atmosphere to undergo a sinusoidal seasonal variation with

an amplitude which ranges sinusoidally from mild to extreme in extent. The variations in the temperature of the polar caps will also cause seasonal variations in the mixing ratio of the volatile atmospheric gases owing to the different behaviours of their saturation vapour pressures with temperature. Triton's visible hemisphere is currently approaching a major southern summer, with solstice scheduled to occur in about 2006 A.D. If the polar caps are not too thin, we should witness a dramatic increase in the CH_4 column abundance before then. . . . Volatiles may exist transiently at lower latitudes, but they are quickly sublimated as the seasonal changes warm them. Consequently, Triton's atmosphere is not saturated at lower latitudes; that is, the lower latitudes may be warmer than the polar latitudes. In this respect, Triton is more like Mars than Pluto. The temperature of the polar caps regulates the mass of the atmosphere. . . . The temperature of Triton's winter polar cap must be as warm as the summer cap, even for the extreme solstices. This is because the surface pressures over the two poles are equal and the surface temperature is specified by the surface pressure when the gases are saturated. . . . One important difference between Triton and Mars is the much greater length of Triton's seasons. If Triton's summer polar cap is sufficiently thin to sublimate entirely during the major summer, the source maintaining Triton's atmosphere would disappear. The sink, however, would still remain; namely, the condensation over the winter polar cap which provides the latent heat to replace the radiation lost to space. Consequently, Triton's atmosphere would commence to freeze out over the winter cap. The extent to which this occurs depends on the volume of the 'volatiles' constituting the cap, as well as on the duration of the seasons. The Martian seasons, for example, are too short for appreciable freezing out of its atmosphere to occur. (Trafton 1984a).

Finally, Trafton made some predictions for the Voyager flyby of August 1989. At a time which was still almost twenty years away from the Neptunian solstice, his model assumed the presence of two polar caps of approximately equal and uniform temperatures; the bulk of the atmosphere was expected to be methane, with some nitrogen, neon, argon, carbon dioxide and oxygen. Both caps would show evidence of shrinking around their borders. For surface regions covered by volatiles, elevated areas would be cooler and depressed areas warmer; the low-flying areas should have the thicker deposits of surface volatiles, because they radiate more heat to space by virtue of their higher temperature (Trafton 1984a). Because of the approach of 'maximum southern summer', a dramatic increase in methane abundance should be observed during the remainder of the century (Trafton 1984b).

Somewhat different ideas have also been proposed by D. P. Cruikshank. The spectral data could be satisfied by a sea of liquid nitrogen covering part of Triton to a depth of a few tens of centimetres; if the condensed nitrogen were present, the atmosphere might be largely N_2, with a pressure of around 0.1 bar and with methane as a minor constituent, while the reddish hue of the surface could be the result of photochemical derivatives of the methane and nitrogen — as with Titan. If methane were dissolved in the liquid nitrogen, red organic matter could be suspended in the liquid.

In summary, the surface of Triton is characterized by solid methane, either as a continuous surface or as icebergs floating in a sea of unknown depth of liquid nitrogen. Reddish photochemical products may give the surface a slight colouration, and water ice may occur as crystals suspended in the liquid nitrogen or as a solid mixed with the methane frosts on expanses of solid surface (spectral modelling favours a suspension of fine crystals in the liquid). In this scenario, the satellite has an atmosphere of nitrogen with other possible minor constituents. (Cruikshank 1984).

It may well be that both condensed methane and condensed nitrogen are present on Triton, so that the most likely surface configuration is an ocean of liquid nitrogen (N_2) with dry areas of solid methane and perhaps some exposed fine-grained water frost — according to J. I. Lunine and D. J. Stevenson (1985). But less than a year before Voyager 2's rendezvous, it was still very uncertain whether the surface of Triton were solid, liquid, or in a condition which could best be described as 'slushy'. It was also claimed that a slight weakening of the methane absorptions in recent years might indicate that the atmosphere had partially clouded or fogged up; there could be a complex seasonal cycle moving Triton's volatiles around enough to ensure that part of the surface always has a light methane frost.

Probably the most dramatic forecast about Triton was made by M. L. Delitsky and W. R. Thompson two years ahead of the Voyager encounter: 'Perhaps Voyager 2, turning its cameras on Triton in 1989, will see plains of white and coloured organic deposits and, maybe, the glint of a distant sun reflected off a calm nitrogen sea' (Delitsky and Thompson 1987).

All the giant planets have satellite families, and initially it did not seem likely that Triton would be Neptune's sole attendant. Lassell, no doubt encouraged by his first success, made a careful search, and once, in 1852 — using a magnification of ×614 — had 'a strong suspicion of a second satellite, but though I saw a point there positively for a short time I could not afterwards see it' (Lassell 1852), and since it was never confirmed we may safely assume that it was either a star or else an optical fault, even though Sir John Herschel, in the 1869 edition of his classic *Outlines of Astronomy*, wrote that 'of the existence of at least two satellites, discovered by Mr. Lassell, there can be no doubt, having also been observed by other astronomers both in Europe and America '— a statement which, despite careful searches of the literature, I have been unable to verify.

J. Schaeberle, using the Lick 36-inch (91-cm) refractor on 24 September 1892, suspected a new satellite — a faint object 24 seconds of arc from Neptune which seemed to change in position angle by two degrees over a period of 1 hour 40 minutes (Schaeberle 1895). This also remained unconfirmed, and a later search by W. H. M. Christie with the 60-inch (152-cm) reflector at Mount Wilson was equally negative. Success finally came in 1949, when G. P. Kuiper detected a second satellite on two plates exposed at the prime focus of the McDonald 82-inch (208-cm) reflector; two pairs of photographs taken during subsequent weeks confirmed that it was indeed a satellite (Kuiper 1949). The magnitude was 19.5, so that it is a very difficult visual object even with giant telescopes. It was named Nereid.

Nereid is small. Its diameter is now known to be only about 240 kilometres; Davies (1988) gave 690 km, with 360 km uncertainty, but this was still rather large to be asteroidal. It has direct motion, but a highly eccentric orbit which takes it from 1.4 to 9.7 million kilometres from Neptune in a period of 359.9 days. This seems more like a cometary orbit than normal satellite motion, and brings us on to the possibility that it has been violently perturbed in the past.

Nereid's faintness makes it very hard to study, but it was found to show quite large-amplitude variations in magnitude. In June 1987 M. W. Schaefer and B. E. Schaefer, using CCD cameras on the 0.9-metre reflector at Cerro Tololo in Chile, made photometric observations showing a range of more than one and a half magnitudes, indicating either an irregular shape or else an unequal albedo over different parts of the surface, as with Saturn's satellite Iapetus (Schaefer and Schaefer 1988). The period was apparently between 8 and 24 hours. It seems that the rotation cannot be synchronous, and indeed synchronous rotation would hardly be expected with such a curious orbit and so great a maximum distance from the primary.

If the variability were due solely to irregularity in shape, the ratio of greatest to least cross-section would be 4 to 1, which seems rather excessive. Even at the most conservative estimate the maximum diameter would have been 338 km and the least 79 km. Almost all satellites with diameters of 200 km or over are approximately spherical, and the investigators concluded that the observed variations were not likely to be due purely to form. If they were due solely to albedo differences, again a ratio of 4 to 1 would be implied, but this would not be out of the question, because the difference in the case of Iapetus is more like 10 to 1. If so, Nereid would have had a diameter of 326 km, with an albedo of 1.0 for the bright side and less than 0.24 for the dark side. In any case, the colour seems to be somewhat unusual. It is possible that Nereid has large-scale surface exposures of segregated and very distinct materials, such as ice and carbonaceous rock (J. Veverka 1988).

As we have seen, searches for rings round Neptune, by the occultation method, led to the suggestion that there might be at least one more satellite detectable in this way — though it was certainly premature to talk about 'the discovery of Neptune's third satellite' (M. Jelečič 1981, S. F. Dermott 1984). But there was every reason to expect that minor satellites might exist, and that the Voyager encounter of August 1989 would reveal some — perhaps even a whole swarm. In any case, it was abundantly clear that the Neptunian system was of immense interest, and quite unlike any other in the known Solar System.

The Voyager 2 results showed that many of the earlier ideas about the satellites had been wrong (see Chapter 13). Triton turned out to be smaller than expected, with a surface coated in part by nitrogen "snow", and with active nitrogen geysers; the extremely tenuous atmosphere proved to be made up chiefly of nitrogen. Nereid was not well imaged by Voyager, and we still do not know a great deal about it. However, six new inner satellites were discovered, and one of these Proteus, is considerably larger than Nereid, though it is so close to Neptune that it is virtually unobservable from Earth.

11

Origin of the Neptunian system

Until long after the discovery of Neptune, the Solar System was assumed to be complete. It was only in 1930 that the ninth planet, Pluto, was found — if indeed Pluto is worthy of planetary rank, which is by no means certain. It was discovered by Clyde Tombaugh as a result of a systematic search carried out at the Lowell Observatory in Flagstaff, Arizona. As the story has been told in detail by the discoverer himself (Tombaugh and Moore 1981) there is no need to repeat it here. Suffice to say that Pluto has proved to be a very curious body. It is a mere 2325 kilometres in diameter, smaller than our Moon; it has a companion, Charon, about 1210 kilometres in diameter, moving in a unique way; the revolution period is equal to Pluto's rotation period (6 d 9 h 17 m) so that the two are 'locked'. The distance between them is no more than 19 000 kilometres. Their surfaces are not alike, since Pluto appears to be covered with methane ice and Charon with water ice, while Pluto has an extremely thin atmosphere, presumably made up chiefly of methane. The orbit is equally unusual. Pluto's distance from the Sun ranges between 4437 million and 7470 million kilometres, so that for part of its orbit it is closer-in than Neptune. This is the case at the present time; perihelion fell in 1989. However, Pluto's orbit is inclined to ours at an angle of 17.14 degrees, and there is no danger of a collision with Neptune at the present epoch (J. G. Benson and G. S. Williams 1971).

But if Pluto, in size and mass, is more like a satellite than a planet — and bearing in mind that it is considerably smaller and less massive than Triton — could it once have been a member of the Neptunian system which broke away, and moved off in an independent orbit? And could this explain the remarkable state of the present Neptunian system, with a large retrograde satellite and a small direct-motion satellite with an orbit of comet-like eccentricity? This was proposed as long ago as 1936 by R. A. Lyttleton. According to his theory, both Pluto and Triton were originally satellites of Neptune, moving in a direct sense; an encounter between them threw Triton into a retrograde orbit, and broke Pluto free altogether, enabling it to masquerade as a planet in its own right (Lyttleton 1936).

Comments were made later by G. P. Kuiper, who maintained that Pluto could not be an original 'protoplanet' because of its unusual orbit (Kuiper 1956). He also stressed that if Pluto were once a satellite of Neptune, it would be expected to have a long rotation period of about a week if its original distance from the planet were less than 60 times Neptune's radius. This value was chosen because Iapetus, Saturn's eighth satellite, is about 60 Saturnian radii out, and has synchronous rotation, so that the same would be expected for all closer-in satellites. (It does not in fact apply to Hyperion, Saturn's seventh satellite, which is closer-in than Iapetus; but this was not known in 1956, and in any case Hyperion is an odd, irregular body with many unusual characteristics.)

Kuiper's paper led to a rather unseemly controversy. W. J. Luyten accused him of merely plagiarizing Lyttleton (Luyten 1956). Kuiper was quick to reply that this was not so, and to stress the difference between his theory and Lyttleton's.

> The conclusion so reached regarding Pluto's origin as a satellite of Neptune should not be confused with Lyttleton's 1936 hypothesis that Pluto and Triton were initially both satellites of Neptune and then had a close encounter which caused Pluto to leave the system and Triton to become retrograde. . . . There is no reason to suppose than an encounter between regular satellites has ever occurred; and there are five retrograde satellites other than Triton. (Kuiper 1957)

Now, thanks to the mutual eclipse phenomena of Pluto and Charon over recent years, we have accurate values for their masses, and W. B. McKinnon has shown that the momentum and energy exchange that would be needed to put Triton into a retrograde path is impossible — so that the simplest hypothesis is that both Triton and Pluto are independent representatives of large outer Solar System planetesimals, in which case Triton was captured by Neptune 'with potentially spectacular consequences that include runaway melting of interior ices and release to the surface of clathrated CH_4, CO and N_2 . . . condensed remnants of this proto-atmosphere could account for features in Triton's unique spectrum' (McKinnon 1984). Also, according to D. N. C. Lin, if the Pluto–Charon system were once a double satellite of Neptune, their combined tidal interaction would have driven them to merge with each other before their orbits around Neptune had had time to evolve significantly; the Pluto–Charon pair is more likely to have been formed by the binary fission of a rapidly rotating object (Lin 1981). The tidally locked state of Pluto and Charon at the present time shows that the total angular momentum is very similar to that which would be required for the break-up of an initially fluid body (F. Mignard 1981).

We must not forget Chiron, discovered by Charles Kowal in 1977, which is officially ranked as an asteroid (No. 2060), although it has some cometary characteristics, but is several hundred kilometres in diameter, assuming a normal albedo, and moves mainly between the orbits of Saturn and Uranus. It could be of the same basic nature as Pluto, though it is much smaller.

Granted that Chiron and the Pluto–Charon pair may be representatives of many other similar objects in the outer Solar System, we could include Triton in the same category, and assume that it was captured by Neptune. If so, it is the largest 'captured body' in the Solar System, with an unusual history and composition, including the possibility of substantial liquid or solid nitrogen content (Stevenson 1984). The capture theory has also been supported on non-dynamical grounds (Čelebonović 1986). And so far as Nereid is concerned, D. P. Cruikshank has commented that 'the inclined, eccentric, direct orbit suggests an origin by capture, in which case Nereid could be representative of the small icy satellites, or could be a dark asteroid of mainly silicate composition' (Cruikshank 1984).

But suppose that we introduce another body — a fairly massive one, presumably a planet, which invaded the outer Solar System in the remote past and caused chaos when it encountered Neptune? This has been discussed by several leading investigators, and although it is admittedly speculative it could at least account for the extraordinary state of affairs which we see today. A single close encounter would produce orbits very similar to those of Pluto, Triton and Nereid. To quote R. S. Harrington and T. C. van Flandern:

> The Neptunian system has probably been disrupted at some unknown time in the past. The most plausible cause of this disruption is an encounter with a massive body. Such an encounter would have produced the observed anomalous features of the orbits of Neptune's satellites, Triton and Nereid. The same event could have inserted a former Neptunian satellite into Pluto's orbit as well, suggesting that Pluto is not unlikely to be an escaped satellite of Neptune.
> . . . Although this planetary body may have escaped the Solar System following the encounter, it is more probable that it has not, and is today an undiscovered planet at a large heliocentric distance. (Harrington and van Flandern 1979)

Moreover, the wanderer could have come close enough to Pluto to have 'ripped off a small chunk that would become a secondary satellite, leaving behind a planet with a very irregular surface and therefore variable reflectivity' (R. S. Harrington and B. J. Harrington 1979). Also, van Flandern has commented that Nereid is 'close to escaping' from Neptune even now; and if Pluto and Charon were originally satellites of Neptune, the one might have captured the other as the disrupting body tore them away (van Flandern 1986).

Against this, S. F. Dermott maintains that 'it is now considered unlikely that Pluto is an escaped satellite. Pluto, Triton and Nereid are probably remnant planetesimals, two of which (at least) were captured by Neptune, destroying in the process any regular satellite system the planet may have possessed' (Dermott 1984).

It must be admitted that the overall situation points to some tremendous event which took place in the remote past. Whether or not Pluto and Charon were involved is not known, but it is at least a distinct possibility. Early in the story of the Solar System, Neptune's satellite system may have been very different from its present state.

New ideas about Pluto were put forward by W. B. McKinnon and S. Mueller in 1988 (McKinnon and Mueller 1988; see also Hughes 1988). New, accurate values for the diameters and masses of Pluto and Charon have become available, thanks to the mutual phenomena of the two, and it appears that the combined mass of the system is about 18.5 per cent that of the Moon. Taken together, they have a density of 1.991, with a low uncertainty (water being taken as unity). If Charon's density is in the region from 1 to 3, as is fairly typical of icy satellites, then Pluto's must be between 1.84 and 2.14, which means that rock makes up from 60 to 80 per cent of its total mass. Moreover, unlike (for instance) Ganymede, its gravitational pressure has not significantly increased the interior rock density, because Pluto is too small. The core rock must be hydrated, because Pluto presumably accreted from silicates and warm water ice, and radioactive decay was inadequate to heat up the core to the 800 to 1200 degrees K needed for differentiation. So either Pluto is undifferentiated, or (more probably) it has a silicate core surrounded by a mantle of ice from 200 to 300 km deep. It cannot have been formed in the inner part of the Solar System — it contains too much ice. The rock-ice ratio is all wrong for an origin in the Jovian satellite system. A collision with an outer asteroid could have broken the original body and produced Charon (and if the original were differentiated, then Charon could have come from the outer part); if this event happened near Neptune, it could account for Pluto's orbital relationship with Neptune. Or Pluto could itself be a large outer-Solar System asteroid which broke up by rotation. In any case, McKinnon and Mueller stress that it is a new kind of object, intermediate between the largest medium-sized icy satellite (Titan) and the smallest icy Galilean satellite (Europa). Comparisons with Triton are of considerable significance.

Quite apart from the possibility of a new, fairly large planet beyond Neptune, to be discussed in Chapter 12, there have been interesting developments since 1992, when David Jewitt and Jane Luu, using the 2.2-metre reflector on Mauna Kea, Hawaii, discovered an asteroidal-sized body with a distance from the Sun varying between 34 and 44 astronomical units, and a period of 296 years. Other similar discoveries followed, and some two dozen of these trans-Neptunian objects have been found by ground-based telescopic searches in the past three years. All of the newcomers so far detected seem to range between about 100 km and 400 km in diameter, although this is uncertain as the reflectivity of their surfaces is not accurately known. They appear to be very dark, reflecting only about four per cent of the sunlight reaching them. It now seems likely that there is a whole swarm of these remote objects, probably members of the hypothetical "Kuiper Belt" named in honour of the late Gerard Kuiper who first suggested its existence. If these icy bodies are regarded as planetesimals, that is to say the "building blocks" from which the main planets were formed in the solar nebula, then it is possible that Pluto may be simply a giant planetesimal — and this could also apply to Triton. In any case, it is clear the outermost part of the Solar System is far less straightforward, and much more interesting, than had been expected.

12

Beyond Neptune

Long before Pluto was discovered, there were various predictions concerning a hypothetical trans-Neptunian planet. G. Dallet, working from the movements of Uranus, proposed a planet at a distance of 47 astronomical units, with a magnitude of between $9\frac{1}{2}$ and $10\frac{1}{2}$ (Dallet 1901). Percival Lowell and W. H. Pickering, independently, also worked upon the perturbations of Uranus, and it was because of the systematic search undertaken at the Lowell Observatory that Pluto was finally found (Tombaugh 1960). But earlier, there had been attempts to predict a new planet from the movements of periodical comets, initially by G. Forbes (1880). Jupiter's comet family was well known, and had been found to contain dozens of members; it was thought that there were less populous families associated with Saturn, Uranus and Neptune — and with the supposed outer planet, which I propose to refer to as Planet X, the popular name for it. T. Grigull studied the orbits of comets seen between 1490 and 1898, and arrived at a planet with a distance of 50.61 astronomical units and a period of 360 years; he even suggested a name for it — Hades (Grigull 1902). T. J. J. See, reasoning along the lines of a somewhat strange 'resisting medium' theory, predicted planets at 42.25, 56 and 72 astronomical units (See 1909, 1910). There was even a Russian amateur, General Alexander Garnowsky, who sent a letter to the Société Astronomique de France about four proposed outer planets. Just what led to this conclusion is not clear (Garnowky 1902).

Initially it was thought that Pluto must be Planet X. In the event, this proved not to be so, because Pluto turned out to be much too lightweight to account for the alleged perturbations — at least of giant worlds such as Uranus and Neptune. It seemed, therefore, that the real Planet X remained to be discovered.

Comets were again called in as evidence. Studies of eight cometary orbits led K. Schütte to assume the existence of a planet at 77 astronomical units (Schütte 1950), and his work was extended by H. H. Kritzinger, whose Planet X moved at 65 astronomical units in a period of $523\frac{1}{2}$ years; the magnitude was assumed to be about 11 (Kritzinger 1954). Later, by 'pairing' data for two of Schütte's eight comets, he amended this distance to 75.1 astronomical units and the period to 650 years, with an inclination of 40 degrees and a magnitude of 10 (Kritzinger 1957). A photographic search was undertaken in the position indicated, but with no result.

Even less convincing was a theory by M. E. Sevin, who believed in a planet moving at 78 astronomical units. His method was to divide the known planets into two groups, inner and outer, and then 'pair' them, but for some obscure reason he included the tiny asteroid 944 Hidalgo, which is only a few kilometres in diameter and has a very eccentric orbit taking it out almost as far as Saturn. 'Pairing' Planet X with Mercury, he produced a planet with a period of 685.8 years, an eccentricity of 0.3, and a mass 11.6 times that of the Earth. Needless to say, no confirmation was forthcoming! (Sevin 1946, Strubell 1952).

Much more recently, the 'comet family' idea has again been discussed, notably by A. S. Guliev and V. P. Tomanov in Russia and by J. J. Matese and D. P. Whitmire. Guliev has claimed that a new cometary family has been found, consisting of comets Halley, di Vico, Westphal, Pons-Gambert, Brorsen-Metcalf and Väisälä 2. Projections of the aphelia of their orbits on to the celestial sphere are concentrated near a large circle, indicating the presence of Planet X moving within the corresponding plane; the distance is given as 36.2 astronomical units (Guliev 1987). More will be said about Halley's Comet shortly. (Westphal's periodical comet appears to have disintegrated; it had a period of 61.9 years, and was quite bright in 1852, but faded out during the return of 1913, and has not been seen since.) Tomanov (1986) came to much the same conclusion as Guliev. J. J. Matese and D. P. Whitmire believe that there are 'showers' of periodical comets associated with the perihelion passages of Planet X through a primordial disk of comets beyond the orbit of Neptune — presumably the same as the hypothetical Oort cloud (Matese and Whitmire 1986a). They also link this with cratering and fossil records showing periods of major impacts on the Earth, modulated with a period of about 30 million years (Matese & Whitmire 1986b). Their Planet X has a distance of between 50 and 100 astronomical units. Finally G. A. Chebotarev, of Leningrad, has studied the aphelia of periodical comets to predict two outer planets, one at 53.7 astronomical units and the other at 100 astronomical units (Chebotarev 1972, 1975).

However, the most spectacular prediction of modern times re-introduced our old friend, Halley's Comet.

More than forty years ago, R. S. Richardson made an attempt to measure the mass of Pluto from its perturbating effects on the comet. He decided that Pluto had no detectable influence, but suggested that there might be an Earth-sized planet moving at a distance of 36.2 astronomical units (approximately 1 astronomical unit beyond the aphelion point of the comet); this would delay the return of the comet to perihelion by one day, while a similar planet at 35.3 astronomical units (0.1 unit beyond Halley's aphelion) would produce a delay of 6 days (Richardson 1942). A somewhat desultory search was put in hand, but with nil result.

Then came the work of J. A. Brady, of the Lawrence Livermore Laboratory of the University of California, in 1972. He maintained that Neptune did show measurable perturbations of unknown origin, but that Halley's Comet was a much more promising subject for investigation. He arrived at a planet at a mean distance of 59.9 astronomical units, with an almost circular orbit (eccentricity 0.07) and a period of 464 years. The orbital inclination, however, was said to be 120 degrees — and the planet was moving in a retrograde sense! The mass was given as three times that of Saturn (Brady 1972). More surprising still, perhaps, the estimated magnitude was between 13 and 14, assuming that the albedo was much the same as Pluto's.

Brady gave a position for his Planet X. It lay, he said, in Cassiopeia, which was frankly about the last place where one would expect to find a major planet. Various searches where made, for example by A. R. Klemoia and E. A. Harlan with the Lick 20-inch double astrograph; they persisted for some time, but found nothing unusual brighter than magnitude 17 to 18 within three degrees of the position given by Brady (Klemola and Harlan 1972). From my own observatory at Selsey, in Sussex, I made a visual and photographic search, though I admit that I had little hope of success; despite the fact that Cassiopeia is a rich region, an object of that magnitude assumed by Brady would have been easy enough to locate — if it had been there. Meanwhile, Clyde Tombaugh had telephoned the Lowell Observatory, and had said, 'I don't think a planet of these characteristics is likely, but it won't hurt to look. These regions have been photographed before, in the early 1940s.' H. Giclas accordingly examined some of Tombaugh's old plates, and found no sign of anything unusual. He then re-photographed the area with the 13-inch refractor which Tombaugh had used to discover Pluto in 1930, again with no result. It had been calculated that Brady's planet, if it existed, would subtend an angle of about 3 seconds of arc, larger than the disk of Neptune (Tombaugh and Moore 1981).

Quite apart from its failure to show up, there were theoretical problems about Brady's planet. P. K. Seidelmann, B. G. Marsden and H. L. Giclas pointed out that it would cause obvious perturbations in the movements of the known planets; it would have an annual proper motion of 4.6 minutes of arc, and it would be too bright to be missed (Seidelmann, Marsden and Giclas 1972). P. Goldreich and W. R. Ward commented that it would make the plane of the Solar System precess in a few tens of millions of years (Goldreich and Ward 1972). Moreover, outgassing jets in the comet itself could account for the delays in returning to perihelion; T. Kiang and P. A. Wayman wrote that Brady's planet was unnecessary, as the residuals in the motions of Halley's Comet could be explained without it (Kiang and Wayman 1973). Before long, it was universally agreed that Brady's planet did not exist; but at least it had directed astronomers' attention back to the problem.

In August 1988 D. Olssen-Steel made the rather surprising suggestion that Neptune was responsible for the capture of Halley's Comet, around 100 000 years ago. He maintained that the comet used to have an orbit taking it from 30 to 100 astronomical units from the Sun, and that it was forced into its present orbit by close encounters with Neptune; the same would be true of the other periodical comets with fairly similar periods (Pons-Gambart, Tempel-Tuttle and Swift-Tuttle). Olssen-Steel presented his paper at the General Assembly of the International Astronomical Union on August 5, but it is fair to say that his conclusions are not generally accepted. (Olssen-Steel 1988).

In September 1988 some new predictions were made by R. Harrington from the U.S. Naval Observatory in Washington. His work was based upon the perturbations of Uranus. His 'planet X' has a period of about 600 years with a mass of from 2–5 times that of the Earth and a present distance of around 9.6 thousand million kilometres. The predicted position is in the region of Scorpius and Sagittarius. It will, of course, be very faint, and whether a search in this area will be successful remains to be seen.

Up to now, the most promising investigations with regard to Planet X have been carried out in the United States by J. D. Anderson, of the Jet Propulsion Laboratory at Pasadena, California. Eschewing Halley's Comet (and all other comets) he has gone back to studies of the perturbations of Uranus and Neptune, and has come to some very remarkable conclusions (Anderson 1986, 1987).

The main point of Anderson's research is that, from all the evidence, there were genuine unexplained perturbations in the movements of Uranus and Neptune between the years 1810 and 1910. They were very slight, but were nevertheless definite. Before 1810, measuring techniques were not accurate enough to produce conclusive evidence one way or the other, and since 1910 the perturbations have apparently ceased. If Planet X were responsible, there is only one explanation. A planet cannot simply 'softly and slightly vanish away', like the hunter of the Snark, so it must travel in a very eccentric orbit — and has now moved so far away that its effects have become inappreciable. Anderson's proposed orbit gives it an inclination of around 90 degrees (that is to say, approximately at right angles to the main plane of the Solar System), with an eccentricity of about 1/3, a mean distance from the Sun of around 80 astronomical units or 7400 million miles (12 000 million kilometres) and a period of from 700 to 1000 years. The mass is given as about 5 times that of the Earth.

There are constraints on the possible mass. Were it is less than 5 times that of the Earth, it would not be adequate to produce the effects of Uranus and Neptune found between 1810 and 1910. Were it much more massive, it would have caused measurable perturbations on the Pioneer 10 and Pioneer 11 space-craft which are now on their way out of the Solar System — and which may, even yet, produce some hard and fast evidence.

Both the Pioneers were Jupiter probes. Pioneer 10 was launched on 2 March 1972, from Cape Canaveral, and encountered Jupiter on 3 December 1973, after having safely negotiated the asteroid belt. It passed Jupiter at 131 400 kilometres, and then began a never-ending journey out of the Solar System.

Pioneer 11 followed a year later; it was launched on 5 April 1973, and passed Jupiter at 46 400 kilometres on 2 December 1974. It had some power left, and, more or less as an afterthought, was swung back across the Solar System to a rendezvous with Saturn on 1 September 1979. Then it, too, began its final journey — but note that it and Pioneer 10 are leaving the Solar System in more or less opposite directions. Contact with them will be maintained for some years yet, and it is hoped that they will keep on sending back signals until they reach the heliopause — that is to say, the region where the solar wind becomes undetectable, and true interstellar space begins.

Up to 1995, no perturbations in the Pioneer motions had been found which could be put down to an unknown planet beyond Neptune and Pluto. This may well rule out any large body moving in a near-circular orbit within reasonable range; in other words, we can probably forget about all the various 'Planet X' theories of conventional type. We can also rule out a more exotic solar companion within this range, such as a 'brown dwarf' (a star which has insufficient mass to spark off nuclear reactions in its core). But a planet with a highly inclined, eccentric orbit is quite another matter, and there is a remote but definite chance that the Pioneers will locate it — if it exists. The two Voyager probes will also leave the Solar System, so that they may give us an extra possibility.

At the moment, this is about as far as we can go. Unless we reject the pre-1910 perturbations of the outer planets, which seems unwise, it is reasonable to assume that Planet X is genuine, but if it is now moving onward toward aphelion in an Anderson-type orbit it will be centuries before it again draws close enough to make the perturbations detectable once more.

Finally, and on a much less serious note, I cannot resist referring to a paper which I published in 1981. In it, I first cast grave doubts upon the validity of Bode's Law, which is not particularly accurate even for some of the naked-eye planets (Saturn, for example) and breaks down completely for Neptune. Pluto was found not so very far from the position given by Lowell. Discounting the idea of coincidence (remember Peirce's 'happy accident' theory!), could it be possible that the real Planet X was in the same part of the sky, but was below the limiting magnitude of Clyde Tombaugh's search? If we assume (a) that Planet X is about as far beyond the orbit of Neptune as that of Neptune is beyond Uranus, (b) that the diameter is much the same as Neptune's and (c) that the orbital inclination and eccentricity are low, we can work out where Planet X is now. With a period of 282 years or so, which fits in with this hypothetical orbit, we can calculate the annual motion since 1930. Pluto was found near the star δ Geminorum, whose right ascension is 7 hours 19 minutes. If the R.A. of Planet X was also 7 h 19 m at that time, the fifty-year interval between 1930 and 1980 would yield an increase of 4 h 15 m, making the R.A. in 1980 11 h 34 m. Round this off to $11\frac{1}{2}$ hours, assume that the inclination is slight, and we reach a position for Planet X near the stars δ and τ Leonis . . . . If the distance and period are slightly greater, we derive a period of 324 years, an increase of 3 h 40 m, and a 1980 value for the right ascension of about 11 h, close to χ Leonis (Moore 1981).

Calculations made on the back of an old envelope, based upon a series of assumptions every one of which is inherently improbable, inspire a feeling of no confidence at all, and I was hardly surprised that my searches with my modest 39-cm reflector led to nil result! But really serious searches have also failed; and all we can say at the moment it that if Planet X exists — as I believe it does — it will be very hard to track down. Only the space-probes can reasonably be expected to provide us with adequate information, and we have to admit that even with the help of the Pioneers and Voyagers we are batting on a very difficult wicket indeed. At the moment, Neptune remains the outermost known planet in the Solar System.

Indeed, the discovery of a swarm of planetesimals, probably belonging to the "Kuiper Belt", may well indicate that, in fact, no more distant large planet beyond the orbit of Neptune actually exists.

Plate 17 – Full disk picture of Neptune taken from a sequence of images acquired by the Voyager 2 narrow-angle camera on 16 and 17 August 1989 showing the Great Dark Spot at latitude 22 degrees south, bright clouds to the south and east of this oval, and a second dark spot at latitude 54 degrees south.

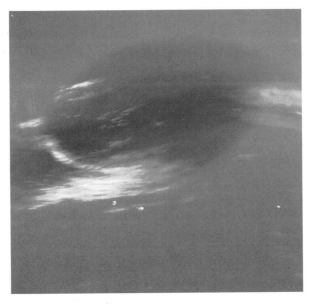

Plate 18 – Close-up image of Neptune's Great Dark Spot taken 45 hours before Voyager 2's closest approach while at a distance of 2.8 million kilometres. The smallest structures visible have a dimension of about 50 kilometres. Feathery white clouds overlie the boundary of the spot, the spiral structure of both the dark boundary and white clouds suggesting a storm system rotating anti-clockwise.

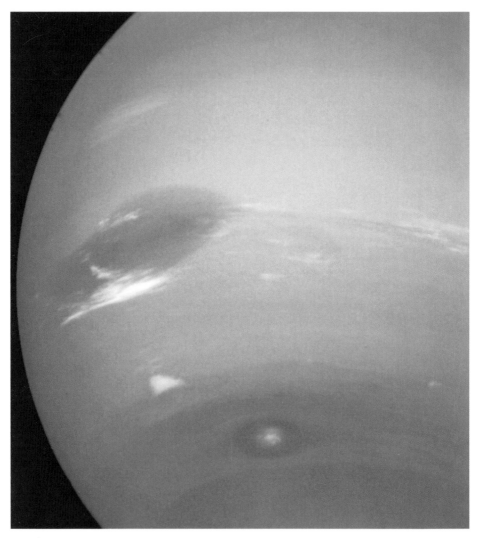

Plate 19 – Voyager 2 image of Neptune showing a number of the planet's cloud features. Towards the north (top) is the Great Dark Spot, accompanied by bright white clouds which undergo rapid changes in appearance. South of the Great Dark Spot is the bright feature nicknamed the "Scooter". Further south is a feature called "Dark Spot 2" which has a bright core. Each feature moves eastward at a different velocity.

Plate 20 – Colour image of Neptune's largest satellite, Triton, obtained by Voyager 2 on 24 August 1989 from a range of 530 000 kilometres. The smallest features shown have a dimension of about 10 kilometres. The surface of Triton is very varied. The most striking feature is the bright southern polar cap, overlying a region known as Uhlanga Regio, which is generally pink in colour. From the ragged edge of the polar cap northwards, Triton's surface is generally darker and redder in colour. Across this darker region roughly parallel to the edge of the cap is a slightly bluish layer possibly due to methane ice crystals scattering the sunlight.

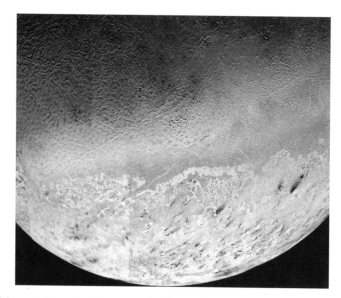

Plate 21 – High resolution colour image mosaic of part of Neptune's satellite, Triton, obtained during Voyager 2's close fly-by on 25 August 1989. The large, pinkish south polar cap (bottom) may consist of a slowly evaporating layer of nitrogen snow and ice, deposited during the previous winter. The dark streaks towards the right of the image are material ejected from the active nitrogen ice geysers erupting from beneath the frozen surface. North of the polar cap, the terrain is smooth and undulating, with walled plains or frozen "lakes" (right), or is characterised by the so-called cantaloupe terrain (left), which resembles a melon skin.

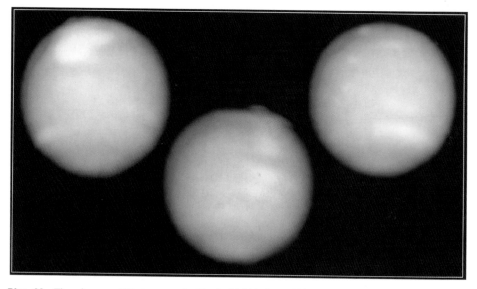

Plate 22 – Three images of Neptune acquired by the Hubble Space Telescope Wide-Field Planetary Camera 2 showing changing weather conditions on the planet. The images were taken in 1994 on October 10 (upper left), October 18 (upper right), and November 2 (lower centre), when Neptune was 4500 million kilometres from Earth. The pink features are high-altitude methane ice crystal clouds, imaged at near-infrared wavelengths. The temperature difference between Neptune's strong internal heat source and its frigid cloud tops may trigger instabilities in the atmosphere that drive the large-scale weather changes observed. Photograph reproduced courtesy of Heidi Hammel (Massachusetts Institute of Technology), NASA, and Space Telescope Science Institute, Baltimore.

13

Epilogue: Voyager to Neptune

All our ideas about the two outer giants, Uranus and Neptune, have been revolutionized by Voyager 2, which passed by Uranus in 1986 and Neptune in 1989. The historical account given in the first twelve chapters of the present book shows how meagre our previous knowledge really was, and the fact that I could list virtually all the important papers in a hundred pages or so demonstrates this. The situation today is very different. Neptune literature is voluminous (see Cruikshank *et al.*, 1995), and this is no place to attempt a list of all the current references, so as an Epilogue I propose to do my best to sum up the new picture of Neptune. I have elsewhere (Moore, 1995) discussed the Voyager missions in more detail.

There were encounters with Jupiter and Saturn by each of two Voyager space-craft, but their trajectories were not identical. Voyager 1 was scheduled to rendezvous first with Jupiter and then with Saturn, but it was also programmed to make a careful survey of Titan, Saturn's largest satellite, which was of special interest because it was known to possess a substantial atmosphere (Gehrels *et al.*, 1984). A pass of Titan would send Voyager 1 well out of the ecliptic plane, so that it would encounter no more planets, and would embark upon a never-ending journey out of the Solar System.

Voyager 2 was also scheduled to rendezvous with Jupiter and Saturn in turn, but its subsequent trajectory depended upon its predecessor. Had Voyager 1 failed to image Titan, then Voyager 2 would have had to do so, and would have sacrificed the passes of Uranus and Neptune. It is therefore understandable that the planners at Mission Control were highly relieved when Voyager 1 functioned perfectly at Titan, leaving its follower free to continue with the exploration of the outer giants.

Voyager 2 was launched first, on 20 August 1977. Voyager 1 followed on 5 September, but followed a more economic orbit, and reached its main targets first: Jupiter on 5 March 1979 and Saturn on 12 November 1980. The images of Titan showed only the upper clouds, but it was established that the atmosphere is considerably denser than that of the Earth, and is made up chiefly of nitrogen, together with a large amount of methane. After the Titan pass, Voyager 1 moved steadily outward; it is still sending back signals, and contact with it should be maintained until well into the 21st century.

Voyager 2 made its rendezvous with Jupiter on 9 July 1979, and encountered Saturn on 25 August 1981. After the closest approach to Saturn there was a crisis when one of the stepper motor and gear train actuators, which control the pointing direction of the scan platform carrying the cameras, jammed and many of the outbound observations were lost; for a time it was feared that the space-craft had been hit by a particle large enough to do serious damage, but the problem subsequently proved to be one of lubrication, and at the Uranus and Neptune encounters the actuator functioned well. It was indeed fortunate that at the time the actuator jammed, the space-craft was in occultation from the Sun by Saturn. An interesting, but little known, point is that a few minutes before the jam, the cameras were pointed at Saturn in a direction which was directly in line with the Sun. If the actuator had stuck at that time, the cameras would have been rendered useless by solar damage after the occultation ended — and there would have been no pictures of Uranus or Neptune.

Voyager 2 surveyed many of Saturn's satellites, but did not make a close approach to Titan; in any case, it could have added little to what Voyager 1 had already told us. Uranus was passed on 24 January 1986, at 80 000 kilometres, and the gravity-assist technique then put Voyager on to a course for the Neptune encounter. It is interesting to note that at this time, NASA estimated the chances of success at Uranus as 60 per cent, but only 40 per cent at Neptune!

Voyager 2 has been described elsewhere (Hunt and Moore, 1994). The space-craft weighed 825 kilogrammes at launch and had a mass of 765 kilogrammes at Neptune (but no weight), the figures decreasing due to propellant consumption. It carries eleven scientific instruments and is dominated by the huge 10-metre magnetometer boom. Solar energy cannot be used at such great distances from the Sun, and so Voyager carries three Radioisotope Thermoelectric Generators (RTGs), which are miniature nuclear power plants. The whole space-craft is amazingly complex, particularly in view of the fact that it had been designed and built in the 1970s, when scientific instrumentation was far less sophisticated than it is now.

The Deep Space Network (DSN) was upgraded for the Neptune pass; each of NASA's three 64-metre tracking antennae was enlarged to 70 metres, a highly efficient 34-metre tracking station was added to the complex at Robledo in Spain, and the 64-metre Parkes radio telescope in Australia was linked with radio equipment at Tidbinbilla. Signals from Voyager were also collected by the Very Large Array (VLA) in New Mexico, and by the 64-metre tracking antenna at Usuda in Japan. The 70-metre and two 34-metre antennae at each DSN complex were linked, as was Parkes linked with those at Tidbinbilla, and the VLA with Goldstone. Usuda signals were combined with Tidbinbilla after the fact. Truly it was an international effort.

The Neptune encounter began officially on 5 June 1989, when Voyager was 117 million kilometres from its target. Closest approach at 29 240 kilometres from the centre of the planet, was at 0356 UT on 25 August; the distance from Earth was then 4427 million kilometres, and the signal transmission time was 4 hours 6 minutes. The "observatory phase" (characterized by repetitive observations of Neptune and its vicinity) extended from 5 June to 6 August; the "far encounter phase" (characterized by unique observations, final instrument calibrations and tests, and final trajectory corrections) from 6 to 24 August; the "near encounter phase" (the closest approach, and highest revolution observations of Neptune and Triton) from 24 to 29 August, and the "post encounter phase" (similar to the observatory and far encounter phases, but on the outbound trajectory) from 29 August to 2 October.

The Voyager planetary programme officially ended on 20 November, but both Voyagers 1 and 2 continue to send back valuable data in a new rôle, searching for evidence of the heliopause, the outermost extent of the Sun's magnetic field and solar wind. The heliopause is where the interstellar medium restricts the outward flow of the solar wind and confines it within a magnetic bubble called the heliosphere. Hopefully the space-craft will still be functioning when they penetrate the heliopause to sample the interstellar medium, allowing measurements to be made of interstellar fields and particles unaffected by the solar plasma.

It had been expected that new small satellites would be found during the Voyager 2 encounter with Neptune, and so it proved; the first of these, now named Proteus, was identified on an image taken in June 1989, and with a diameter of 416 kilometres is actually larger than Nereid, though its closeness to Neptune makes it virtually impossible to observe from Earth. Five others followed, all within 80 000 kilometres of Neptune and all below 200 kilometres in diameter; they were subsequently named Naiad, Thalassa, Despina, Galatea and Larissa. Proteus and Larissa were imaged. Larissa was almost certainly responsible for an occultation event seen in 1981, and then attributed to a ring-arc; it is darkish, with a cratered surface. Proteus was better shown, and found to be somewhat squarish in shape. On the Neptune-facing side surveyed by Voyager 2 there is one major feature, the Southern Hemisphere Depression, which is almost 250 kilometres in diameter and 10 kilometres deep, with a rugged floor; there are also troughs and ridges, plus several well-marked craters, one of which is over 80 kilometres across.

The other new satellites were not imaged in detail, but there is no reason to doubt that they, too, are darkish and cratered. One disappointment was that Nereid was in an unfavourable part of its comet-like orbit, and Voyager never approached it closer than 4.7 million kilometres, so that no definite surface detail could be made out. It is unlikely that its rotation period is equal to its orbital period of 360.16 days, so that its rotation, like that of Hyperion in Saturn's system, is chaotic.

The rings proved to be of great interest, though they are much less prominent even than those of Uranus. The theory of incomplete ring arcs (see Chapter 9) proved to be wrong; the rings are continuous, and are four in number, plus what is termed the "Plateau". Four of the main rings have been named after astronomers who were closely concerned with the discovery of Neptune; I have made the official suggestion that the Plateau should be named in honour of D'Arrest. Details follow overleaf.

The Plateau consists of a diffuse band of material containing a high percentage of very small particles; there may also be 'dust' going all the way down to the outer clouds, though there seems to be a relatively empty region just inside the Le Verrier ring. The Arago ring is close to the orbit of the newly-discovered satellite Galatea, which may be in the nature of a "shepherd", though efforts to find other small satellites within the ring system have so far been unsuccessful. It is the Adams ring which is the most significant. It is somewhat reddish, and in this respect is similar to the obscure ring of Jupiter; it is "clumpy", with three brighter segments – and it is these which gave rise to the pre-Voyager suggestion that there might have been incomplete arcs rather than full rings. During the Voyager pass, the Adams ring occulted the second-magnitude star Sigma Sagittarii, and this showed that the ring

Name	Distance from centre of Neptune, km	Width, km	Notes
Galle	41 900	1 700	High dust content
Le Verrier	53 200	<50	High dust content
Plateau	53 200 – 59 000	5 800	Diffuse band
Arago	62 000	<30	Orbit of Galatea
Adams	62 900	<50	Contains three brighter arcs

The Galle and Le Verrier rings contain a great deal of dust.

has a core only 17 kilometres wide. The three arcs are clustered within a range of 33 degrees in longitude.†

The immense difficulty of obtaining images of the rings is shown by the fact that exposures of almost 600 seconds were needed to obtain some of the pictures. Even in the Adams ring, particles occupy only about 10 per cent of the total space, so that the rings are very flimsy compared even with those of Uranus.

Voyager crossed the plane of the rings at a velocity of over 76 000 kilometres per hour. Impacts were recorded 40 minutes before the crossing, and reached a peak of 300 per second for 10 to 15 minutes to either side of the actual crossing. It was a great relief when the space-craft emerged unharmed.

It had been assumed that Neptune would have a magnetic field, and so it proved; radio emissions were detected at a fairly early stage in the encounter, but there was a delay before Voyager 2 met the bow shock, which is located at 879 000 kilometres from the planet. The major surprise was that, as with Uranus, the magnetic axis is nowhere near the axis of rotation; the angle between them is 47 degrees, as against 58.6 degrees for Uranus – and with Neptune, too, the magnetic axis does not pass through the centre of the globe; it is displaced by 10 000 kilometres. The rotation period of the magnetic field, as measured from the periodicity of the radio emissions, is 16.1 hours, which corresponds to the rotation period of the deep interior of the planet; therefore all the meteorological observations are measured relative to this period.

The field strength is 1.2 gauss in the southern hemisphere, but only 0.06 gauss in the northern, so that Neptune's magnetic field is weaker than those of the other giants, and the magnetosphere is relatively "empty" even though it extends out to over 25 Neptune radii. The dynamo electric currents are presumably closer to the surface than to the deep core, but it is still not clear why Neptune, which has a "normal" rotational tilt, has a magnetic axis of the same type as that of Uranus. We can discount

---

† At the 1991 General Assembly of the International Astronomical Union, held in Buenos Aires, it was decided to name the three arcs in the Adams ring, Liberté, Egalité and Fraternité. This is a departure from the accepted schemes of nomenclature, and with its somewhat political undertones appears to be rather undesirable. Whether the names will come into general use is doubtful. It would have been far more appropriate to have named them in honour of astronomers such as Bouvard, Bessel and Hussey.

the idea that both giants are going through magnetic reversals, because this would surely be too much of a coincidence.

The rings, plus all the satellites apart from Nereid, move within the Neptunian magnetosphere; and as the magnetic field rotates with the planet in 16.1 hours, the satellites and ring particles are swept through the highly-charged region, which is bound to affect their basic properties and surface chemistry.

Lightning was detected from charged particle discharges studied by the plasma wave equipment, though it was not observed visually. Auroræ were confirmed, but are of course most intense near the magnetic poles rather than near the poles of rotation. The auroral power is weak, amounting to only about 5×10^7 watts as against 10^{11} watts for terrestrial auroræ.

As Voyager closed in, Neptune's beautiful blue surface became more and more striking; we were obviously dealing with a world far more dynamic than the bland Uranus. The two giants may have been twins in size and mass, but they were proving to be very much in the nature of non-identical twins.

The most conspicuous surface feature on Neptune at the time of the Voyager 2 encounter was a huge oval, the Great Dark Spot, south of the equator. It was about the same size as the Earth; its size, relative to Neptune, was about the same as that of the Great Red Spot relative to Jupiter. Of course it was not red, but was slightly "less blue" than its surroundings. It had a rotation period of 18.296 hours, and drifted westward at about 30 metres per second relative to the adjacent clouds; it rotated in an anti-clockwise direction, and showed more or less predictable changes in aspect ratio. It was about 10 per cent darker than its surroundings, while the nearby material was 30 per cent brighter — indicative of the differences in altitude between the two regions; the Great Dark Spot had all the characteristics of an atmospheric vortex. Hanging above it were bright clouds which gave the impression of being cirrus, made up of methane ice; between the cirrus and the main cloud-deck there was a clear region about 50 kilometres deep. The cirrus changes shape quite quickly, and in some cases there are shadows cast on the cloud-deck beneath — a phenomenon not observed on Jupiter or Saturn, and certainly not on Uranus.

Further south, at latitude –42°, there is a smaller, very variable feature which speeds round Neptune in a period of 15.97 hours, and has therefore been nicknamed "the Scooter". It has a bright centre, and shows rapid variations in shape; every few revolutions it catches up to the Great Dark Spot and passes it. Still further south (latitude –55°,) there is another dark spot, D2. It also has a bright core, and shows small-scale features inside it which change markedly over periods of a few hours.

There are violent winds on Neptune. At the equator they blow westward (retrograde) at up to 450 metres per second; further south they slacken, and beyond latitutde –50° they become eastward at up to 300 metres per second, decreasing once more near the south pole. In fact, a broad equatorial retrograde jet extends from approximately +45° to –50°, with a relatively narrow prograde jet at around –70°, latitude. Neptune is therefore the "windiest" planet in the Solar System; as the heat budget is only 1/20 of that at Jupiter, it seems that the winds are so strong because of the relative lack of turbulence.

Temperature measurements show that there is a cold mid-latitude region, with a warmer equator and pole. The temperature is much the same as that of Uranus; at the equator, –226°C as against –214°C for Uranus. This is because Neptune has an excess energy balance of 2.61, indicating a strong internal heat source, while the inner heat source of Uranus is negligible and may even be lacking altogether. Obviously we are still unsure about Neptune's internal structure, but it seems likely there is a core, out to 0.2 of the total radius, which may be essentially silicate; this is surrounded by the mantle, but the core does not seem to be strongly differentiated from the icy components (such as water) which dominate the globe.

At least we have positive information about the nature of the atmosphere, and here again the fortunate occultation of Sigma Sagittarii proved to be very useful indeed. The upper atmosphere consists of 85 per cent hydrogen, 13 per cent helium and only 2 per cent methane. At a level where the pressure is 3.3 bars there is a layer which seems to be made up of hydrogen sulphide, and above this come layers of hydrocarbons such as ethylene and acetylene, with a methane layer and upper methane haze. Above the hydrogen sulphide layer there are discrete clouds with diameters of the order of 100 kilometres, and it is these which cast shadows on the cloud-deck 50 to 75 kilometres below, though the methane cirrus clouds straight above the Great Dark Spot and spot D2 were not observed to throw shadows.

Apparently there is a definite cycle of events. First, solar ultra-violet destroys methane high in Neptune's atmosphere by converting it to other hydrocarbons such as ethane and acetylene. These hydrocarbons sink to the lower stratosphere, where they evaporate and condense. The hydrocarbon ice particles fall into the warmer troposphere, where they evaporate and are converted back to methane. Buoyant, convective methane clouds then rise up to the base of the stratosphere or higher, returning methane vapour to the stratosphere and preventing any net methane loss. In the troposphere we can expect variable amounts of hydrogen sulphide, methane and ammonia, all of which are involved in the creation of the cloud layers and associated photochemical processes.

As Voyager skimmed over Neptune's darkened north pole, at 96 000 kilometres per hour, no pictures could be obtained; quite apart from the lack of sunlight, the motion was too fast and any pictures would have been hopelessly smeared. Next came the outbound crossing of the ring-plane, and then just over five hours later, Voyager encountered its last target — Triton, which provided a fitting climax to an extraordinary mission.

It had already become clear that many previous ideas had been wrong; there were no oceans of any sort, and the atmosphere was extremely rarefied, so that there was no analogy with Titan. Moreover, Triton was smaller than expected — smaller than the Moon, with a diameter of only 2705 kilometres, though this still made it larger than Pluto. The surface temperature proved to be −235°C, so that Triton is the coldest body ever encountered by a space-craft. The globe is fairly dense, with a specific gravity of 2.06, so that the material is probably about two-thirds rock and one-third ice. The escape velocity is 1.44 kilometres per second — less than that of the Moon, though Triton's much lower temperature means that, unlike the Moon, it can retain an atmosphere; the ground atmospheric pressure is of the order of 14 microbars, which is 70 000 times less than that of the Earth's air. The main constituent is nitrogen, in the form of N_2, which accounts for 99 per cent; most of the rest is methane, with a trace of carbon monoxide. There is considerable haze, seen by Voyager above the surface, and is probably composed of microscopic ice crystals of methane or nitrogen. Winds in the atmosphere average around 5 metres per second westward, though naturally they have very little force. There is a pronounced temperature inversion, since the temperature in the atmosphere rises to about −173°C at a height of 600 kilometres. This inversion occurs at a surprisingly high altitude, for reasons which are unclear.

The surface of Triton is very varied. There is a general coating of ice, presumably water ice, overlaid by nitrogen and methane ices; water ice has not been detected spectroscopically, but it must exist, because nitrogen and methane ices are not hard enough to maintain surface relief over long periods. Not that there is much surface relief on Triton; there are no mountains or deep valleys, and the total surface relief cannot amount to more than a few hundred metres. Normal craters are scarce; the largest is a mere 27 kilometres in diameter. Extensive flows are seen, some of them up to 80 kilometres wide, due probably to ammonia-water fluids.

The most striking feature is the southern polar cap, which is covered with pink nitrogen snow and ice. The areas surveyed by Voyager have been divided into three regions: Bubembe Regio (western equatorial), Monad Regio (eastern equatorial) and Uhlanga Regio (polar). These names, and those of other features, have been allotted by the Nomenclature Committee of the International Astronomical Union.

Uhlanga is covered by the pink cap, with some of the underlying geological units showing through in places. It is here that we find the nitogen ice geysers which were so unexpected. Apparently there is a layer of liquid nitrogen from twenty to thirty metres below the surface; here the pressure is high enough for the nitrogen to remain liquid, but if for any reason it migrates upward it will come to a region where the pressure is only about 1/10 that of the Earth's air at sea-level. The nitrogen will then explode in a shower of ice and vapour (about 80 per cent ice, 20 per cent vapour) and will travel up the nozzle of the geyser-like vent at a rate of up to 150 metres per second, which is fast enough to make it rise to several kilometres before falling back. The outrush sweeps dark débris along it, producing plumes of material such as Viviane Macula and Namazu Macula, from 15 to 50 kilometres wide and 50 to 75 kilometres long.

The edge of the cap, separating Uhlanga from Monad and Bubembe, is sharp and convoluted. The long Tritonian seasons mean that the southern pole has been in constant sunlight for over a century now, and along the cap borders there are signs of evaporation; the cap may eventually appear to shift northward, so that the entire aspect of Triton may change according to its seasons. In reality it is not the cap shifting northward, but evaporation/sublimation products moving northward in the atmosphere where at the colder norther regions they are redeposited to form the raw materials for the processes here to be repeated during the northern summer. North of the cap there is a darker, redder region; the colour may be due to the action of solar ultra-violet upon the methane. Running across this region, more or less parallel to the edge of the cap, is a slightly bluish layer, due possible to tiny crystals of methane ice scattering the incoming sunlight.

Monad Regio is part smooth, with knobbly or hummocky terrain; there are rimless pits (paterae) with graben-like troughs (fossae) and strange, mushroom-like features (guttae), such as Zin and Akupara, whose origin is unclear. There are also walled plains or "lakes", such as Tuonela and Ruach, edged with terraces as though the original level has been changed several times by repeated melting and re-freezing. The interiors of these lakes are flat and smooth. Undulating smooth plains cover much of the higher part of Monad.

Bubembe Regio is characterized by the so-called cantaloupe terrain — a nickname give to it because of a superficial resemblance to a melon skin! Fissures cross it, meeting in high X or Y junctions. Liquid material, presumably a mixture of ammonia and water, seems to have forced its way up some of these fissures, so that there are central ridges; material has even flowed out on to the plain before freezing there. The cantaloupe areas are probably the oldest parts of Triton's surface; the pateræ are interpreted as explosive features of the caldera type, while the pits could well be due to collapse.

Quite apart from the nitrogen geysers, Triton's surface must be variable on a much larger scale. Southern midsummer will not fall until around the year 2006, and there will no doubt be marked changes in the cap — and also in the north polar region, which is at the moment plunged into its long winter night.

It seems highly probable that Triton has not always been a satellite of Neptune, but was once an independent body which was subsequently captured by Neptune. Its initial orbit around the planet would have been eccentric, but over a period of perhaps 1000 million years the path would have been forced into the present circular form; for a time there may have been a dense atmosphere, and certainly the interior will have been tidally flexed and heated, accompanied by marked surface activity which in the long run produced the strange scene which was revealed by Voyager.

The lifetime of features in the atmosphere of Neptune, particularly the Great Dark Spot, could not be determined from the Voyager 2 data alone, because the space-craft encounter was too brief, and these features cannot be resolved with ground-based telescopes. Fortunately, the Hubble Space Telescope (HST) is now able to resolve the disk of Neptune almost as well as most ground-based telescopes can resolve the disk of Jupiter. HST images of Neptune obtained in June and October 1994 showed no sign of either the Great Dark Spot, or its smaller companion D2. However, the images revealed many bright clouds in Neptune's atmosphere, including a prominent cloud band in the northern hemisphere centred near latitude 30°N. Northern hemisphere clouds were less obvious at the time of the Voyager 2 flyby.

Interestingly, in November 1994, HST discovered a new great dark spot, this time located in Neptune's northern hemisphere. The new feature appeared near the limb of the planet because the northern hemisphere of Neptune was then tilted away from Earth. The new spot is a near mirror-image of the southern hemisphere dark spot that was discovered by Voyager. Like its predecessor, the new spot is accompanied by bright high-altitude clouds along its edge, probably formed of methane ice crystals. Such dramatic changes on Neptune are not fully understood, but they emphasize the dynamic nature of the planet's atmosphere.

For further detailed close-range information about Neptune we must await a new space-probe. In the present political and economic climate, we may have to wait for a long time. Voyager 2 can tell us no more; by 29 August 1989 it was already 7 million kilometres from Neptune, and the blue planet had become a tiny disk, while Triton had shrunk to a mere dot of light. Voyager will not return; millions of years hence it may still be moving through interstellar space — unseen, unheard and untraceable.

We do not yet claim to have a complete knowledge of Neptune, but at least we have learned more than would have seemed possible before the flight of Voyager 2. Certainly we have come a long way since Heinrich D'Arrest called out "That star is not on the map!" a hundred and fifty years ago.

Appendix 1: Data

1. NEPTUNE

Distance from Sun, 10^6km (a.u.): mean		4496.7 (30.058)
	maximum	4537 (30.316)
	minimum	4456 (29.800)

Sidereal period, years (days)	164.8 (60 190.3)
Rotation period	16 h 7 m
Mean orbital velocity, km/s	5.43
Axial inclination	28°48′
Orbital inclination	1°45′19″.8
Orbital eccentricity	0.009

Diameter, km: equatorial	50 538
polar	49 600
Apparent diameter from Earth	max. 2″.2, min. 2″.0
Reciprocal mass, Sun = 1	19 300
Density, water = 1	1.77
Mass, Earth = 1	17.2
Volume, Earth = 1	57
Escape velocity, km/s	23.9
Surface gravity, Earth = 1	1.2
Mean surface temperature	−220°C
Oblateness	0.02
Albedo	0.35
Maximum magnitude	+7.7
Mean diameter of Sun, seen from Neptune	1′04″

2.   NEPTUNE'S SATELLITES

Satellite	Mean distance from centre of Neptune, km	Orbital period days	(hours)	Orbital inclination, degrees	Orbital eccentricity	Diameter, km	Magnitude	Albedo
Naiad	48 200	0.296	(7.1)	4.5	0	54 ± 16	25	0.06
Thalassa	50 000	0.3112	(7.5)	0	0	80 ± 16	24	0.06
Despina	52 500	0.333	(8.0)	0	0	180 ± 20	24	0.06
Galatea	62 000	0.429	(110.3)	0	0	150 ± 30	24	0.054
Larissa	73 600	0.554	(13.3)	0	0	192	21	0.056
Proteus	117 600	1.121	(26.9)	0	0	416	20	0.050
Triton	354 800	5.877		159.9	0.0002	2705	13.6	0.6–0.8
Nereid	5 514 000	360.16		27.2	0.749	240	18.7	0.16

Triton has a mass of 2.14×10^{22} kg, a density of 2.06 kg/m^3, and an escape velocity of 1.44 km/s.
Nereid's distance from Neptune ranges between 1 345 500 and 9 688 500 km.
Before the new satellites were named, they were given provisional designations: 1989 N1 (Proteus), N2 (Larissa), N3 (Despina), N4 (Galatea), N5 (Thalassa) and N6 (Naiad).
Triton was discovered by Lassell in 1846, and Nereid by Kuiper in 1949.

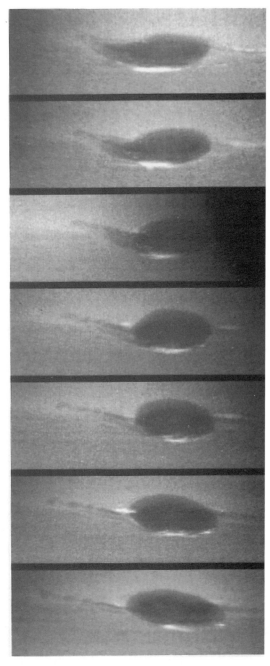

Plate 23 – Sequence of Voyager 2 images showing changes in the clouds around Neptune's Great Dark Spot over a four-and-a-half day period. From top to bottom the images show successive rotations of the planet. The sequence shows a large change in the western end (left side) of the Spot, where a dark extension seen in the earlier images converges into an extended 'string' of small dark spots over the next five rotations.

Plate 24 – Two 590-second exposures of the rings of Neptune acquired by the Voyager 2 wide-angle camera on 26 August 1989 from a range of 280 000 kilometres. The two main rings (Adams, outer and Le Verrier, inner) are clearly visible and are complete all the way round the planet. (The brighter segments in the outer Adams ring were not visible in either exposure.) Also shown are the inner faint Galle ring and the diffuse band or 'Plateau' between the two main rings.

Plate 25 – Voyager 2 wide-angle image, taken after closest encounter, showing Neptune's rings in detail. The two main rings are the inner Le Verrier ring, 53 200 kilometres from the centre of Neptune and the outer Adams ring, 62 900 kilometres distant, which contains three brighter segments or ring arcs.

(a)

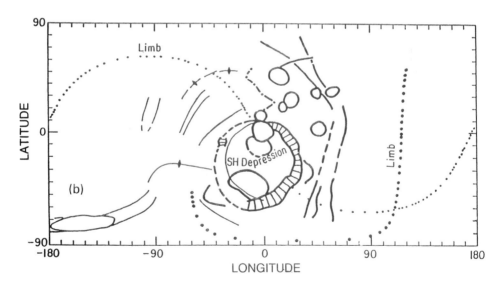

(b)

Plate 26 – (a) Voyager 2 image of Neptune's second largest satellite, Proteus, obtained on August 25 1989, from a range of 146 000 kilometres. The satellite, seen here about half illuminated, has a mean diameter of 416 kilometres. It is rather dark, with an albedo of about 6 percent, and grey in colour. (b) Map showing the principal surface features on Neptune's satellite, Proteus. There is one major feature, the Southern Hemisphere Depression, which is nearly 250 kilometres across and 10 kilometres deep, with a rugged floor. There are many troughs and ridges, plus several well-marked craters, one of which is more than 80 kilometres across.

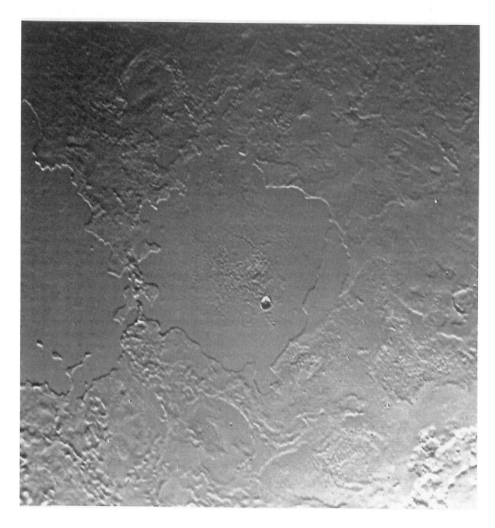

Plate 27 – Part of Neptune's satellite, Triton, within the region known as Monad Regio, showing two walled plains or frozen "lakes". These are possibly old impact basins, that have been extensively modified by flooding, melting, faulting, and collapse. Several episodes of filling and partial removal of material appear to have occurred.

Plate 28 – Part of Neptune's satellite, Triton, within the region known as Bubembe Regio, which is characterised by the so-called canteloupe terrain, a nickname given to it because of a superficial resemblance to a melon skin! Fissures criss-cross the terrain, meeting in huge X or Y junctions. Liquid material, presumably a mixture of ammonia and water ice, seems to have welled up along some of these fissures so that there are central ridges.

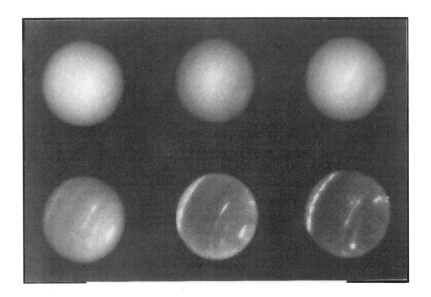

Plate 29 – A sequence of Hubble Space Telescope images of Neptune, taken through a series of filters, obtained in June 1994. Top row, left to right: filter passband centred near wavelength 300 nm (near-ultraviolet); 467 nm (blue); and 588 nm (green). Bottom row, left to right: filters centred near 673 nm (red); 619 nm (red methane band); 889 nm (infrared methane band). The bright clouds in Neptune's atmosphere are most obvious at the red and infrared wavelengths which methane gas absorbs. Images reproduced courtesy of J. Trauger *et al.*, NASA/Jet Propulsion Laboratory, and Space Telescope Science

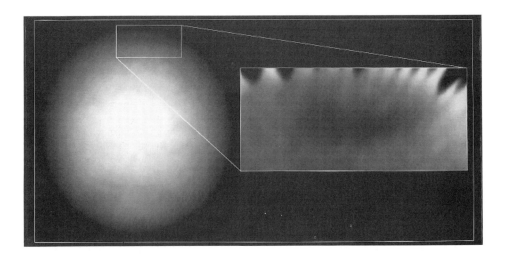

Plate 30 – Hubble Space Telescope image of Neptune acquired with the Wide-Field Planetary Camera 2 on 2 November 1994, showing a new dark spot in Neptune's northern hemisphere. Because the planet's northern hemisphere is now tilted away from Earth, the new spot appears near the limb of the planet. Earlier in 1994, the Hubble Space Telescope showed that the Great Dark Spot, seen by Voyager 2 in Neptune's southern hemisphere, had disappeared. Photograph reproduced courtesy of Heidi Hammel (Massachusetts Institute of Technology), NASA, and Space Telescope Science Institute, Baltimore.

Appendix 2:
Airy's 'Account', 13 November 1846

ROYAL ASTRONOMICAL SOCIETY

VOL. VII.	*Nov.* 13, 1846.	No. 9.

CAPTAIN W. H. SMYTH, R.M., President, in the Chair.

John Riddle, Esq., Second Master of the Nautical School, Greenwich Hospital, was ballotted for and duly elected a Fellow of the Society.

The Astronomer Royal read the following Memoir:—

I. Account of some circumstances historically connected with the discovery of the Planet exterior to *Uranus*.

IT has not been usual to admit into the *Memoirs* of this Society mere historical statements of circumstances which have occurred in our own times. I am not aware that this is a matter of positive regulation: it is, I believe, merely a rule of practice, of which the application in every particular instance has been determined by the discretion of those Officers of the Society with whom the arrangement of our *Memoirs* has principally rested. And there can be no doubt that the ordinary rule must be a rule for the exclusion of papers of this character; and that if a positive regulation is to be made, it must absolutely forbid the presentation of such histories. Yet it is conceivable that events may occur in which this rule ought to be relaxed; and such, I am persuaded, are the circumstances attending the discovery of the planet exterior to *Uranus*. In the whole history of astronomy, I had almost said in the whole history of science, there is nothing comparable to this. The history of the discoveries of new planets in the latter part of the last century, and in the present century, offers nothing analogous to it. *Uranus, Ceres,* and *Pallas*, were discovered in the course of researches which did not contemplate the possible discovery of planets. *Juno* and *Vesta*, were discovered in following up a series of observations suggested by a theory which, fruitful as it has been, we may almost venture to call fanciful. *Astræa* was found in the course of a well-conducted re-examination of the heavens, apparently contemplating the discovery of a new planet as only one of many possible results. But the motions of *Uranus*, examined by philosophers who were fully impressed with the universality of the law of gravitation, have long exhibited the effects of some disturbing body: mathematicians have at length ventured on the task of ascertaining where such a body could be; they have pointed out that the supposition of a disturbing body moving in a certain orbit, precisely indicated by them, would entirely explain the observed disturbances of *Uranus:* they have expressed their conviction, with a firmness which I must characterise as wonderful, that the disturbing planet would be found exactly in a certain spot, and presenting exactly a certain appearance; and in that spot, and with that appearance, the planet has been found. Nothing in the whole history of astronomy can be compared with this.

The principal steps in the theoretical investigations have been made by one individual, and the published discovery of the planet was necessarily made by one individual. To these persons the public attention has been principally directed; and well do they deserve the honours which they have received, and which they will continue to receive. Yet we should do wrong if we considered that these two persons alone are to be regarded as the authors of the discovery of this planet. I am confident that it will be found that the discovery is a consequence of what may properly be called a movement of the age; that it has been urged by the feeling of the scientific world in general, and has been nearly perfected by the collateral, but independent labours, of various persons possessing the talents or powers best suited to the different parts of the researches.

With this conviction, it has appeared to me very desirable that the authentic history of this discovery should be published as soon as possible; not only because it will prove a valuable contribution to the history of science, but also because it may tend to do justice to some persons who otherwise would not receive in future times the credit which they deserve. And as a portion of the history, I venture to offer to this Society a statement of the circumstances which have come to my own knowledge. I have thought that I could with propriety do this: not because I can pretend to know all the history of the discovery, but because I know a considerable part of it; and because I can lay claim to the character of impartiality to this extent, that, though partaking of the general movement of the age, I have not directly contributed either to the theoretical or to the observing parts of the discovery. In a matter of this delicacy I have thought it best to act on my own judgment, without consulting any other person; I have, however, solicited the permission of my English correspondents for the publication of letters.

Without pretending to fix upon a time when the conviction of the irreconcilability of the motions of *Uranus* with the law of gravitation first fixed itself in the minds of some individuals, we may without hesitation date the general belief in this irreconcilability from the publication of M. Alexis Bouvard's *Tables of Uranus* in 1821. It was fully shewn in the introduction to the tables, that, when every correction for perturbation indicated by the best existing theories was applied, it was still impossible to reconcile the observations of Flamsteed, Lemonnier, Bradley, and Mayer, with the orbit required by the observations made after 1781: and the elements of the orbit were adopted from the latter observations, leaving the discordances with the former (amounting sometimes to three minutes of arc) for future explanation.

The orbit thus adopted represented pretty well the observations made in the years immediately following the publication of the tables. But in five or six years the discordance again growing up became so great, that it could not escape notice. A small error was shewn by the Kremsmünster Observations of 1825 and 1826: but, perhaps, I am not in error in stating that the discordance was first prominently exhibited in the Cambridge Observations, the publication of which from 1828 was conducted under my superintendance.

While still residing at Cambridge, I received from the Rev. T. J. Hussey (now Dr. Hussey) a letter, of which the following is an extract. It will be considered, I think, as honourable to that gentleman's acuteness and zeal. I must premise that the writer had lately passed through Paris.

No. 1.   *The Rev. T. J. Hussey to G. B. Airy.*
[EXTRACT.]

"Hayes, Kent, 17 *November,* 1834.

"With M. Alexis Bouvard I had some conversation upon a subject I had often meditated, which will probably interest you, and your opinion may determine mine. Having taken great pains last year with some observations of *Uranus,* I was led to examine closely Bouvard's tables of that planet. The apparently inexplicable discrepancies between the ancient and modern observations suggested to me the possibility of some disturbing body beyond *Uranus,* not taken into account because unknown. My first idea was to ascertain some approximate place of this supposed body empirically, and then with my large reflector set to work to examine all the minute stars thereabouts: but I found myself totally inadequate to the former part of the task. If I could have done it formerly, it was beyond me now, even supposing I had the time, which was not the case. I therefore relinquished the matter altogether; but subsequently, in conversation with Bouvard, I inquired if the above might not be the case: his answer was, that, as might have been expected, it had occurred to him, and some correspondence had taken place between Hansen and himself respecting it. Hansen's opinion was, that one disturbing body would not satisfy the phenomena; but that he conjectured there were two planets beyond *Uranus.* Upon my speaking of obtaining the places empirically, and then sweeping closely to the bodies, he fully acquiesced in the propriety of it, intimating that the previous calculations would be more laborious than difficult; that if he had leisure he would undertake them and transmit the results to me, as the basis of a very close and accurate sweep. I have not heard from him since on the subject, and have been too ill to write. What is your opinion on the subject? If you consider the idea as possible, can you give me the limits, roughly, between which this body or those bodies may probably be found during the ensuing winter? As we might expect an excentricity [inclination?] approaching rather to that of the old planets than of the new, the breadth of the Zone to be examined will be comparatively inconsiderable. I may be wrong, but I am disposed to think that, such is the perfection of my equatorial's object-glass, I could distinguish, almost at once, the difference of light of a small planet and a star. My plan of proceeding, however, would be very different: I should accurately map the whole space within the required limits, down to the minutest star I could discern; the interval of a single week would then enable me to ascertain any change. If the whole of this matter do not appear to you a chimæra, which, until my conversation with Bouvard, I was afraid it might, I shall be very glad of any sort of hint respecting it."

My answer was in the following terms:—

No. 2.   *G. B. Airy to the Rev. T. J. Hussey.*
[EXTRACT.]

"Observatory, Cambridge, 1834, *Nov.* 23.

"I have often thought of the irregularity of *Uranus,* and since the receipt of your letter have looked more carefully to it. It is a puzzling subject, but I give it as my opinion, without hesitation, that it is not yet in such a state as to give the smallest hope of making out the nature of any external action on the planet. Flamsteed's observations I reject (for the present) without ceremony: but the two observations by Bradley and Mayer cannot be rejected. Thus the state of things is this,—the mean motion and other elements derived from the observations between 1781 and 1825 give considerable errors in 1750, and give *nearly the same errors in* 1834, *when the planet is at nearly the same part of its orbit.* If the mean motion had been determined by 1750 and 1834, this would have indicated nothing: but the fact is, that the mean motions were determined (as I have said) independently. This does not look like irregular perturbation. The observations would be well reconciled if we could from theory bring in two terms; one a small error in Bouvard's excentricity and perihelion, the other a term depending on twice the longitude. The former, of course, we could do; of the latter there are two, viz. a term in the equation of the centre, and a term in the perturbations by *Saturn.* The first I have verified completely (formula and numbers); the second I have verified generally, but not completely: I shall, when I have the opportunity, look at it thoroughly. So much for my doubts as to the certainty of any extraneous action. But if we are certain that there were any extraneous action, I doubt much the possibility of determining the place of the planet which produced it. I am sure it could not be done till the nature of the irregularity was well determined from several successive revolutions."

It will readily be understood that I do not quote this letter as a testimony to my own sagacity; but I think it deserving of production, as shewing the struggle which was made twelve years ago to explain the motions of *Uranus*, and the difficulty which seemed to envelope the subject. With regard to my last sentence, I think it likely that the same difficulty would still have been felt, if the theorists who entered seriously upon the explanation of the perturbations had not trusted more confidently to Bode's law of distances than I did myself.

In the year 1836, having quitted the Observatory of Cambridge, I completed the reduction of the planetary observations made there during the years 1833, 1834, 1835, in such a form as to exhibit the heliocentric errors of the tabular places of *Uranus*, together with the effect of errors of the tabular radius vector. The memoir containing these reductions was subsequently printed in the *Memoirs* of this Society. The progress of the errors of the tables of *Uranus* was here clearly marked.

In 1837, I received from M. Eugène Bouvard a letter, from which I trust I may be permitted to make an extract. It will, I am certain, be received as creditable to the intelligence and industry of the astronomers of the Observatory of Paris.

No. 3.   *M. Eugène Bouvard to G. B. Airy*
[EXTRACT.]

"Paris, ce 6 Octobre, 1837.

"Dans le peu de moments de loisir que me laissent mes fonctions, je m'occupe d'un travail que je crois n'être pas sans importance. Mon oncle [M. Alexis Bouvard] travaille à réfaire ses tables de *Jupiter* et de *Saturne*, en se servant des corrections apportées recemment aux élémens astronomiques. Il m'a cédé les tables d'*Uranus* a réconstruire. En consultant les comparaisons que vous avez fait des observations de cette planète avec les calculs des tables, on voit que les différences en latitude sont très-grandes et qu'elles vont toujours en augmentant. Cela tient-il à une perturbation inconnue apportée dans les mouvemens de cet astre par un corps situé au-delà? Je ne sais, mais c'est de moins l'idée de mon oncle. Je regarde la solution de cette question comme fort importante. Mais, pour réussir, j'ai besoin de réduire les observations avec la plus grande précision, et souvant les moyens me manquent."

The remainder of this letter relates principally to the reduction of observations. The following are extracts from my answer:—

No. 4.   G. B. Airy to M. Eugène Bouvard.
[EXTRACT.]
"Royal Observatory, Greenwich, 1837, Oct. 12.

"I think that, probably, you would gain much in the accuracy of the reduced observations by waiting a short time before you proceed with that part of your labour. Some time ago, I presented to the Astronomical Society of London a very complete reduction of the observations of all the planets made at Cambridge in the years 1833, 1834, 1835. This paper will, as I expect, very shortly be printed. I have reduced the observations made at Greenwich in 1836 in the same manner: the volume containing these reductions will very soon be published. * * * You may also know that I am engaged upon a general reduction of the observations of planets made at Greenwich, from the commencement of Bradley's observations to the present time. It may, perhaps, be a year before I can furnish you with the places deduced from these observations. * * * With respect to the errors of tables of *Uranus*, I think you will find that it is the *longitude* which is most defective and that the errors in *latitude* are not at present increasing. To shew this, I set down a few of my results. * * * You will see by this statement that the errors of longitude are increasing with fearful rapidity, while those of latitude are nearly stationary. * * * I cannot conjecture what is the cause of these errors, but I am inclined, in the first instance, to ascribe them to some error in the perturbations. There is no error in the pure elliptic theory (as I found by examination some time ago). If it be the effect of any unseen body, it will be nearly impossible ever to find out its place."

On the 24th of February, 1838, I addressed a letter to M. Schumacher, which is printed in the *Astronomische Nachrichten,* No. 349. In this letter it is shewn, by treatment of the results of the reduced observations of 1833, 1834, 1835, 1836 (to which allusion was made in my letter to M. Eugène Bouvard), that the tabular radius vector of *Uranus* was considerably too small. This deduction (which has been confirmed by the observations of all the subsequent years) has always appeared to me to be very important. It is, perhaps, worth while here to point out that the detection of this error arose, in the first place, from the circumstance that my observations of *Uranus* had not been confined to the mere opposition (as had too often been done), but had been extended, as far as possible, to quadratures; and, in the next place, from my having so reduced the observations as to exhibit the effect of error of the radius vector.

On the 14th of May, 1838, I transmitted to M. Eugène Bouvard the reduced observations of 1833, 1834, 1835, 1836; and referred him to a paper in the *Astronomische Nachrichten* which I have cited.

The following letter from M. Eugène Bouvard will, shew how vigorously the attention of the astronomers of Paris was still directed to *Uranus:*—

No. 5.   M. Eugène Bouvard to G. B. Airy.
[EXTRACT.]
"Paris, ce 21 Mai, 1844.

"* * * Je viens aujourd'hui vous prier de me communiquer, si c'est possible, les ascensions droites et les déclinaisons d'*Uranus* depuis 1781 jusqu'en 1800. * * * J'ai réduit moi-même toutes ces observations en m'en tenant aux élémens imprimés, mais je crains qu'il n'y ait quelques erreurs. Il y a surtout une telle incertitude sur les erreurs de collimation de quart de cercle depuis 1785 jusqu'en 1800, qu'il est presque impossible d'avoir une grande confiance dans les observations. * * * Mon travail est fort avancé. Je suis arrivé à des resultats fort bons déjà, puisque je satisfais aux observations actuelles et aux premières de 1781, 1782, &c., à 15″ de dégré près en longitude: tandisque d'après les tables de mon oncle les erreurs sont de près de 2′ de dégré actuellement. Si je mettais de côté les observations de Maskelyne faites depuis 1785 jusqu'à 1796, mes tables pourraient satisfaire aux observations à 7″ ou 8″ près. Mais je crains que cette période ne m'empêche d'y parvenir; et malheureusement c'est dans cette intervalle que les observations sont le plus défectueuses. * * * D'après mes calculs, il faut changer considérablement les élémens elliptiques d'Herschel, surtout le moyen mouvement et le périhélie. J'ai determiné aussi la masse de *Saturne*, et je la trouve très differente de celle que l'on admet; il faut l'augmenter beaucoup. Mais j'attendrai une nouvelle approximation pour être tout à fait sûr de ma détermination."

After some further correspondence, I transmitted to M. Eugène Bouvard, on June 27, 1844, the proof-sheets of the *Planetary Reductions*, containing the Right Ascensions and North Polar distances of *Uranus;* and M. Bouvard, in acknowledging the receipt of them, on July 1, 1844, pointed out an error in the refraction for June 15, 1819. I mention this to shew the extreme care with which M. E. Bouvard's collateral calculations had been conducted.

Although no allusion is made in the last letter to the possible disturbing planet, it would be wrong to suppose that there was no thought of it. In fact, during the whole of these efforts for reforming the tables of *Uranus*, the dominant thought was, "Is it possible to explain the motions of *Uranus*, without admitting either a departure from the received law of attraction, or the existence of a disturbing planet?" I know not how far the extensive and accurate calculations of M. Eugène Bouvard may have been used in the subsequent French calculations, but I have no doubt whatever that the knowledge of the efforts of M. Bouvard, the confidence in the accuracy of his calculations, and the perception of his failure to reconcile in a satisfactory way the theory and the observations, have tended greatly to impress upon astronomers, both French and English, the absolute necessity of seeking some external cause of disturbance.

I have departed from a strictly chronological order for the sake of keeping in connexion the papers which relate to the same trains of investigation. Several months before the date of the last letter quoted, I had received the first intimation of those calculations which have led to a distinct indication of the place where the disturbing planet ought to be sought. The date of the following letter is Feb. 13, 1844:—

No. 6. *Professor Challis to G. B. Airy.*
[EXTRACT.]

"Cambridge Observatory, Feb. 13, 1844.

"A young friend of mine, Mr. Adams, of St. John's College, is working at the theory of *Uranus*, and is desirous of obtaining errors of the tabular geocentric longitudes of this planet, when near opposition, in the years 1818–1826, with the factors for reducing them to errors of heliocentric longitude. Are your reductions of the planetary observations so far advanced that you could furnish these data? and is the request one which you have any objection to comply with? If Mr. Adams may be favoured in this respect, he is further desirous of knowing, whether in the calculation of the tabular errors any alterations have been made in Bouvard's *Tables of Uranus* besides that of *Jupiter's* mass."

My answer was as follows:—

No. 7. *G. B. Airy to Professor Challis.*
[EXTRACT.]
"Royal Observatory, Greenwich, 1844, *Feb.* 15.

"I send all the results of the observations of *Uranus* made with both instruments [that is, the heliocentric errors of *Uranus* in longitude and latitude from 1754 to 1830, for all those days on which there were observations, both of right ascension and of polar distance]. No alteration is made in Bouvard's *Tables of Uranus*, except increasing the two equations which depend on *Jupiter* by $\frac{1}{50}$ part. As constants have been added (in the printed tables) to make the equations positive, and as $\frac{1}{50}$ part of the numbers in the tables has been added, $\frac{1}{-50}$ part of the constants has been subtracted from the final results."

Professor Challis, in acknowledging the receipt of these, used the following expressions:—

No. 8. *Professor Challis to G. B. Airy.*
[EXTRACT.]
"Cambridge Observatory, Feb. 16, 1844.

"I am exceedingly obliged by your sending so complete a series of tabular errors of *Uranus.* * * * The list you have sent will give Mr. Adams the means of carrying on in the most effective manner the inquiry in which he is engaged."

The next letter shews that Mr. Adams had derived results from these errors.

No. 9. *Professor Challis to G. B. Airy.*
"Cambridge Observatory, Sept. 22, 1845.

"My friend Mr. Adams (who will probably deliver this note to you) has completed his calculations respecting the perturbation of the orbit of *Uranus* by a supposed ulterior planet, and has arrived at results which he would be glad to communicate to you personally, if you could spare him a few moments of your valuable time. His calculations are founded on the observations you were so good as to furnish him with some time ago; and from his character as a mathematician, and his practice in calculation, I should consider the deductions from his premises to be made in a trustworthy manner. If he should not have the good fortune to see you at Greenwich, he hopes to be allowed to write to you on this subject."

On the day on which this letter was dated, I was present at a meeting of the French Institute. I acknowledged it by the following letter:—

No. 10. *G. B. Airy to Professor Challis.*
"Royal Observatory, Greenwich, 1845, *Sept.* 29.

"I was, I suppose, on my way from France, when Mr. Adams called here: at all events, I had not reached home, and therefore, to my regret, I have not seen him. Would you mention to Mr. Adams that I am very much interested with the subject of his investigations, and that I should be delighted to hear of them by letter from him?"

On one of the last days of October, 1845, Mr. Adams called at the Royal Observatory, Greenwich, in my absence, and left the following important paper:—

No. 11.   *J. C. Adams, Esq. to G. B. Airy.*

"According to my calculations, the observed irregularities in the motion of *Uranus* may be accounted for by supposing the existence of an exterior planet, the mass and orbit of which are as follows:—

Mean Distance (assumed nearly in accordance with Bode's law).	38·4
Mean Sidereal Motion in 365·25 days	1°30′·9
Mean Longitude, 1st October, 1845	323 34
Longitude of Perihelion	315 55
Excentricity	0·1610.
Mass (that of the Sun being unity), ...	0·0001656.

For the modern observations I have used the method of normal places, taking the mean of the tabular errors, as given by observations near three consecutive oppositions, to correspond with the mean of the times; and the Greenwich observations have been used down to 1830: since which, the Cambridge and Greenwich observations and those given in the *Astronomische Nachrichten*, have been made use of. The following are the remaining errors of mean longitude:—

Observation — Theory.

	″		″		″
1780	+0·27	1801	−0·04	1822	+0·30
1783	−0·23	1804	+1·76	1825	+1·92
1786	−0·96	1807	−0·21	1828	+2·25
1789	+1·82	1810	+0·56	1831	−1·06
1792	−0·91	1813	−0·94	1834	−1·44
1795	+0·09	1816	−0·31	1837	−1·62
1798	−0·99	1819	−2·00	1840	+1·73

The error for 1780 is concluded from that of 1781 given by observation, compared with those of four or five following years, and also with Lemonnier's observations in 1769 and 1771.

"For the ancient observations, the following are the remaining errors:—

Observation — Theory.

	″		″		″
1690	+44·4	1750	−1·6	1763	− 5·1
1712	+ 6·7	1753	+5·7	1769	+ 0·6
1715	− 6·8	1756	−4·0	1771	+11·8

The errors are small, except for Flamsteed's observation of 1690. This being an isolated observation, very distant from the rest, I thought it best not to use it in forming the equations of condition. It is not improbable, however, that this error might be destroyed by a small change in the assumed mean motion of the planet."

I acknowledged the receipt of this paper in the following terms:—

No.12.   *G. B. Airy to J. C. Adams, Esq.*

"*Royal Observatory, Greenwich,* 1845. *Nov.* 5.

"I am very much obliged by the paper of results which you left here a few days since, shewing the perturbations on the place of *Uranus* produced by a planet with certain assumed elements. The latter numbers are all extremely satisfactory: I am not enough acquainted with Flamsteed's observations about 1690 to say whether they bear such an error, but I think it extremely probable.

"But I should be very glad to know whether this assumed perturbation will explain the error of the radius vector of *Uranus*. This error is now very considerable, as you will be able to ascertain by comparing the normal equations, given in the Greenwich observations for each year, for the times *before* opposition with the times *after* opposition."

I have before stated, that I considered the establishment of this error of the radius vector of *Uranus* to be a very important determination. I therefore considered that the trial, whether the error of radius vector would be explained by the same theory which explained the error of longitude, would be truly an *experimentum crucis*. And I waited with much anxiety for Mr. Adams's answer to my query. Had it been in the affirmative, I should at once have exerted all the influence which I might possess, either directly, or indirectly through my friend Professor Challis, to procure the publication of Mr. Adams's theory.†

From some cause with which I am unacquainted, probably an accidental one, I received no immediate answer to this inquiry. I regret this deeply, for many reasons.

While I was expecting more complete information on Mr. Adams's theory, the results of a new and most important investigation reached me from another quarter. In the *Compte Rendu* of the French Academy for the 10th November, 1845, which arrived in this country in December, there is a paper by M. Le Verrier on the perturbations of *Uranus* produced by *Jupiter* and *Saturn*, and on the errors in the elliptic elements of *Uranus*, consequent on the use of erroneous perturbations in the treatment of the observations. It is impossible for me here to enter into details as to the conclusions of this valuable memoir: I shall only say that, while the correctness of the former theories, as far as they went, was generally established, many small terms were added; that the accuracy of the calculations was established by duplicate investigations, following different courses, and executed with extraordinary labour; that the corrections to the elements, produced by treating the former observations with these corrected perturbations, were obtained; and that the correction to the ephemeris for the present time, produced by the introduction of the new perturbations and the new elements, was investigated, and found to be incapable of explaining the observed irregularity of *Uranus*. Perhaps it may be truly said that the theory of *Uranus* was now, for the first time, placed on a satisfactory foundation. This important labour, as M. Le Verrier states, was undertaken at the urgent request of M. Arago.

In the *Compte Rendu* for June 1, 1846, M. Le Verrier gave his second memoir on the theory of *Uranus*. The first part contains the results of a new reduction of nearly all the existing observations of *Uranus*, and their treatment with reference to the theory of perturbations, as amended in the former memoir. After concluding from this reduction that the observations are absolutely irreconcilable with the theory, M. Le Verrier considers in the second part all the possible explanations of the discordance, and concludes that none is admissible, except that of a disturbing planet exterior to *Uranus*. He then proceeds to investigate the elements of the orbit of such a planet, assuming that its mean distance is double that of *Uranus*, and that its orbit is in the plane of the ecliptic. The value of the mean distance, it is to be remarked, is not fixed entirely by Bode's law, although suggested by it; several considerations are stated which compel us to take a mean distance, not *very* greatly differing from that

---

† Here the Astronomer Royal explained to the meeting, by means of a diagram, the nature of the errors of the tabular radius vector.

suggested by the law, but which nevertheless, without the suggestions of that law, would leave the mean distance in a most troublesome uncertainty. The peculiarity of the form which the investigation takes is then explained. Finally, M. Le Verrier gives us the most probable result of his investigations, that the true longitude of the disturbing planet for the beginning of 1847 must be about 325°, and that an error of 10° in this place is not probable. No elements of the orbit or mass of the planet are given.

This memoir reached me about the 23rd or 24th of June. I cannot sufficiently express the feeling of delight and satisfaction which I received fom it. The place which it assigned to the disturbing planet was the same, to one degree, as that given by Mr. Adams's calculations, which I had perused seven months earlier. To this time I had considered that there was still room for doubt of the accuracy of Mr. Adams's investigations; for I think that the results of algebraic and numerical computations, so long and so complicated as those of an inverse problem of perturbations, are liable to many risks of error in the details of the process: I know that there are important numerical errors in the *Mécanique Céleste* of Laplace; in the *Théorie de la Lune* of Plana; above all, in Bouvard's first tables of *Jupiter* and *Saturn*; and to express it in a word, I have always considered the correctness of a distant mathematical result to be a subject rather of moral than of mathematical evidence. But now I felt no doubt of the accuracy of both calculations, as applied to the perturbation in longitude. I was, however, still desirous, as before, of learning whether the perturbation in radius vector was fully explained. I therefore addressed to M. Le Verrier the following letter:—

No. 13.   *G. B. Airy to M. Le Verrier.*
"Royal Observatory, Greenwich, 1846, June 26.

"I have read with very great interest, the account of your investigations on the probable place of a planet disturbing the motions of *Uranus*, which is contained in the *Compte Rendu de l'Académie* of June 1; and I now beg leave to trouble you with the following question. It appears, from all the later observations of *Uranus* made at Greenwich (which are most completely reduced in the *Greenwich Observations* of each year, so as to exhibit the effect of an error either in the tabular heliocentric longitude, or the tabular radius vector), that the tabular radius vector is considerably too small. And I wish to inquire of you whether this would be a consequence of the disturbance produced by an exterior planet, now in the position which you have indicated?

"I imagine that it would not be so, because the principal term of the inequality would probably by analogous to the Moon's variation, or would depend on sin 2 $(v-v')$; and in that case the perturbation in radius vector would have the sign — for the present relative position of the planet and *Uranus*. But this analogy is worth little, until it is supported by proper symbolical computations.

"By the earliest opportunity I shall have the honour of transmitting to you a copy of the *Planetary Reductions*, in which you will find all the observations made at Greenwich to 1830 carefully reduced and compared with the tables."

Before I could receive M. Le Verrier's answer, a transaction occurred which had some influence on the conduct of English astronomers.

On the 29th June, a meeting of the Board of Visitors of the Royal Observatory of Greenwich was held, for the consideration of special business. At this meeting, Sir J. Herschel and Professor Challis (among other members of the Board) were present; I was also present, by invitation of the Board. The discussion led, incidentally, to the general question of the advantage of distributing subjects of observation among different observatories. I spoke strongly in favour of such distribution; and I produced, as an instance, the extreme probability of now discovering a new planet in a very short time, provided the powers of one observatory could be directed to the search for it. I gave, as the reason upon which this probability was based, the very close coincidence between the results of Mr. Adams's and M. Le Verrier's investigations of the place of the supposed planet disturbing *Uranus*. I am authorised by Sir J. Herschel's printed statement in the *Athenæum* of October 3, to ascribe to the strong expressions which I then used the remarkable sentence in Sir J. Herschel's address, on September 10, to the British Association assembled at Southampton. "We see it [the probable new planet] as Columbus saw America from the shores of Spain. Its movements have been felt, trembling along the far-reaching line of our analysis, with a certainty hardly inferior to that of ocular demonstration."†. And I am authorised by Professor Challis, in oral conversation, to state that the same expressions of mine induced him to contemplate the search for the suspected planet.

M. Le Verrier's answer reached me, I believe, on the 1st of July. The following are extracts from it:—

No. 14.   *M. Le Verrier to G. B. Airy.*
[EXTRACT.]

"Paris, 28 *Juin,* 1846

"* * * Il a toujours été dans mon désir de vous en écrire, aussi qu'à votre savante Société. Mais j'attendais, pour cela, que mes recherches fussent complètes, et ainsi moins indignes de vous être offertes. Je compte avoir terminé la rectification des éléments de la planète troublante avant l'opposition qui va arriver; et parvenir à connaître ainsi les positions du nouvel astre avec une grande précision. Si je pouvais espérer que vous aurez assez de confiance dans mon travail pour chercher cette planète dans le ciel, je m'empresserais, Monsieur, de vous envoyer sa position exacte, dès que je l'aurai obtenue.

"La comparaison des positions d'*Uranus*, observées dans ces dernières années, dans les oppositions et dans les quadratures, montre que le rayon de la planète, calculé par les tables en usage, est effectivement très-inexact. Cela n'a pas lieu dans mon orbite, telle que je l'ai déterminée; il n'y a pas plus d'erreur dans les quadratures que dans les oppositions.

"Le rayon est donc bien calculé dans mon orbite; et, si je ne me trompe, M. Airy désirerait savoir quelle est la nature de la correction que j'ai fait subir à cet égard aux tables en usage?

"Vous avez raison, Monsieur, de penser que cette correction n'est pas due à la perturbation du rayon vecteur produite *actuellement* par la planète troublante. Pour s'en rendre un compte exact, il faut remarquer que l'orbite d'*Uranus* a été calculée par M. Bouvard sur des positions de la planète qui n'étaient pas *les positions elliptiques*, puisqu'on n'avait pas pu avoir égard aux perturbations produites par la planète inconnue. Cette circonstance a nécessairement rendu les éléments de l'ellipse faux, et c'est à l'erreur de l'excentricité et à l'erreur de la longitude de périhélie qu'il faut attribuer l'erreur actuelle du rayon vecteur d'*Uranus*.

"Il résulte de ma théorie que l'excentricité donnée par M. Bouvard doit être augmentée, et qu'il en est de même de la longitude du périhélie; deux causes qui

---

† This sentence is copied from the written draft of the speech. Sir J. Herschel appeared to suppose that the sentence had not been reported in the public journals as spoken. I did, however, see it so reported in an English newspaper, to which I had access on the Continent.

contribuent, à cause de la position actuelle de la planète dan son orbite, à augmenter le rayon vecteur. Je ne transcris pas ici les valeurs de ces accroissements, parceque je ne les ai pas encore avec toute la rigueur précise, mais je les aurai rectifié avant un mois, et je me ferai un devoir. Monsieur, de vous les transmettre aussitôt, si cela vous est agréable.

"Je me bornerai à ajouter que la position en quadrature, déduite en 1844 des deux oppositions qui la comprennent, au moyen de mes formules, ne diffère de la position observée que de $0''\cdot6$; ce qui prouve que l'erreur du rayon vecteur est entièrement disparue.

"C'est même une des considérations qui devront donner plus de probabilité à la vérité de mes résultats, qu'ils rendent un compte scrupuleux de toutes les circonstances du problème. Ainsi, bien que je n'aye fait usage dans mes premières recherches que des oppositions, les quadratures n'ont pas laissé de se trouver calculées avec toute l'exactitude possible. Le rayon vecteur s'est trouvé rectifié de lui-même, sans que l'on l'eut pris en considération d'une manière directe. Excusez-moi, Monsieur, d'insister sur ce point. C'est une suite de désir que j'ai d'obtenir votre suffrage.

"Je recevrai avec bien de plaisir les observations que vous voulez bien m'annoncer. Malheureusement le temps presse; l'opposition approche; il faut de toute necessité que j'aye fini pour cette époque. Je ne pourrai donc pas comprendre ces observations dans mon travail. Mais elles me seront très-utiles pour me servir de vérifications; et c'est ce à quoi je les employerai certainement."

It is impossible, I think, to read this letter without being struck with its clearness of explanation, with the writer's extraordinary command, not only of the physical theories of perturbation but also of the geometrical theories of the deduction of orbits from observation, and with his perception that his theory *ought* to explain all the phenomena, and his firm belief that it *had* done so. I had now no longer any doubt upon the reality and general exactness of the prediction of the planet's place. My approaching departure for the Continent made it useless for me to trouble M. Le Verrier with a request for the more accurate numbers to which he alludes; but the following correspondence will shew how deeply his remarks had penetrated my mind.

About a week after the receipt of M. Le Verrier's letter, while on a visit to my friend the Dean of Ely, I wrote to Professor Challis as follows:—

No. 15. *G. B. Airy to Professor Challis.*

"The Deanery, Ely, 1846, July 9.

"You know that I attach importance to the examination of that part of the heavens in which there is * * * * reason for suspecting the existence of a planet exterior to *Uranus*. I have thought about the way of making such examination, but I am convinced that (for various reasons, of declination, latitude of place, feebleness of light, and regularity of superintendence) there is no prospect whatever of its being made with any chance of success, except with the Northumberland Telescope.

"Now I should be glad to ask you, in the first place, whether you could make such an examination?

"Presuming that your answer would be in the negative, I would ask, secondly, whether, supposing that an assistant were supplied to you for this purpose, you would superintend the examination?

"You will readily perceive that all this is in the most unformed state at present, and that I am asking these questions almost at a venture, in the hope of rescuing the matter from a state which is, without the assistance that you and your instruments can give, almost desperate. Therefore I should be glad to have your answer, not only responding simply to my questions, but also entering into any other considerations which you think likely to bear on the matter.

"The time for the said examination is approaching near."

In explanation of this letter, it may be necessary to state that, in common, I believe, with other astronomers at that time, I thought it likely that the planet would be visible only in large telescopes. I knew that the Observatory of Cambridge was at this time oppressed with work, and I thought that the undertaking—a survey of such an extent as this seemed likely to prove—would be entirely beyond the powers of its personal establishment. Had Professor Challis assented to my proposal of assistance, I was prepared immediately to place at his disposal the services of an efficient assistant; and for approval of such a step, and for liquidation of the expense which must thus be thrown on the Royal Observatory, I should have referred to a Government which I have never known to be illiberal when demands for the benefit of science were made by persons whose character and position offered a guarantee, that the assistance was fairly asked for science, and that the money would be managed with fair frugality. In the very improbable event of the Government refusing such indemnity, I was prepared to take all consequences on myself.

On the 13th July, I transmitted to Professor Challis "Suggestions for the Examination of a Portion of the Heavens in search of the external Planet which is presumed to exist and to produce disturbance in the motion of *Uranus*," and I accompanied them with the following letter:—

No. 16.　　*G. B. Airy to Professor Challis.*
"Royal Observatory, Greenwich, 1846, *July,* 13.

"I have drawn up the enclosed paper, in order to give you a notion of the extent of work incidental to a sweep for the possible planet.

"I only add at present that, in my opinion, the importance of the inquiry exceeds that of any current work, which is of such a nature as not to be totally lost by delay."

My "Suggestions" contemplated the examination of a part of the heavens 30° long, in the direction of the ecliptic, and 10° broad. They entered into considerable details as to the method which I proposed; details which were necessary, in order to form an estimate of the number of hours' work likely to be employed in the sweep.

I received, in a few days, the following answer:—

No. 17.　　*Professor Challis to G. B. Airy.*
[EXTRACT.]
"Cambridge Observatory, July 18*th,* 1846.

"I have only just returned from my excursion. * * * I have determined on sweeping for this hypothetical planet. * * * With respect to your proposal of supplying an assistant I need not say any thing, as I understand it to be made on the supposition that I decline undertaking the search myself. * * * I purpose to carry the sweep to the extent you recommend."

The remainder of the letter was principally occupied with the details of a plan of observing different from mine, and of which the advantage was fully proved in the practical observation.

On August 7, Professor Challis, writing to my confidential assistant (Mr. Main) in my supposed absence, said,—

No. 18.   *Professor Challis to the Rev. R. Main.*
[EXTRACT.]

"Cambridge Observatory, August 7, 1846.

"I have undertaken to search for the supposed new planet more distant than *Uranus*. Already I have made trial of two different methods of observing. In one method, recommended by Mr. Airy * * * I met with a difficulty which I had anticipated. * * * I adopted a second method."

From a subsequent letter (to be cited hereafter), it appears that Professor Challis has commenced the search on July 29, and had actually observed the planet on August 4, 1846.

Mr. Main's answer to the other parts of this letter, written by my direction, is dated August 8.

At Wiesbaden (which place I left on September 7), I received the following letter from Professor Challis:—

No. 19.   *Professor Challis to G. B. Airy.*
[EXTRACT.]

"Cambridge Observatory, Sept. 2, 1846.

"I have lost no opportunity of searching for the planet; and, the nights having been generally pretty good, I have taken a considerable number of observations: but I get over the ground very slowly, thinking it right to include all stars to 10–11 magnitude; and I find, that to scrutinise, thoroughly, in this way the proposed portion of the heavens, will require many more observations than I can take this year."

On the same day on which Professor Challis wrote this letter, Mr. Adams, who was not aware of my absence from England, addressed the following very important letter to Greenwich:—

No. 20.   *J. C. Adams, Esq. to G. B. Airy.*
"St. John's College, Cambridge, Sept. 2, 1846.

"In the investigation, the results of which I communicated to you last October, the mean distance of the supposed disturbing planet is assumed to be twice that of *Uranus*. Some assumption is necessary in the first instance, and Bode's law renders it probable that the above distance is not very remote from the truth: but the investigation could scarcely be considered satisfactory while based on any thing arbitrary; and I therefore determined to repeat the calculation, making a different hypothesis as to the mean distance. The eccentricity also resulting from my former calculations was far too large to be probable; and I found that, although the agreement between theory and observation continued very satisfactory down to 1840, the difference in subsequent years was becoming very sensible, and I hoped that these errors, as well as the eccentricity, might be diminished by taking a different mean distance. Not to make too violent a change, I assumed this distance to be less than the former value by about $\frac{1}{20}$th part of the whole. The result is very satisfactory, and appears to shew that, by still further diminishing the distance, the agreement between the theory and the later observations may be rendered complete, and the eccentricity reduced at the same time to a very small quantity. The mass and the elements of the orbit of the supposed planet, which result from the two hypotheses, are as follows:—

	Hypothesis I. $\left(\dfrac{a}{a^1} = 0.5\right)$	Hypothesis II. $\left(\dfrac{a}{a^1} = 0.515\right)$
Mean Longitude of Planet, 1st Oct. 1846	325° 8′	323° 2′
Longitude of Perihelion.........................	315 57	299 11
Eccentricity	0·16103	0·12062
Mass (that of Sun being 1).....................	0·00016563	0·00015003

"The investigation has been conducted in the same manner in both cases, so that the differences between the two sets of elements may be considered as wholly due to the variation of the fundamental hypothesis. The following table exhibits the differences between the theory and the observations which were used as the basis of calculation. The quantities given are the errors of *mean* longitude, which I found it more convenient to employ in my investigations than those of the true longitude.

Ancient Observations.

Date.	(Obs.—Theory.) Hypoth I.	Hypoth. II.	Date.	(Obs.—Theory.) Hypoth I.	Hypoth. II.
	″	″		″	″
1712	+6·7	+6·3	1756	− 4·0	− 4·0
1715	+6·8	−6·6	1764	− 5·1	− 4·1
1750	−1·6	−2·6	1769	+ 0·6	+ 1·8
1753	+5·7	+5·2	1771	+11·8	+12·8

Modern Observations.

	″	″		″	″
1780	+0·27	+0·54	1810	+0·56	+0·61
1783	−0·23	−0·21	1813	−0·94	−1·00
1786	−0·96	−1·10	1816	−0·31	−0·46
1789	+1·82	+1·63	1819	−2·00	−2·19
1792	−0·91	−1·06	1822	+0·30	+0·14
1795	+0·09	+0·04	1825	+1·92	+1·87
1798	−0·99	−0·93	1828	+2·25	+2·35
1801	−0·04	+0·11	1831	−1·06	−0·82
1804	+1·76	+1·94	1834	−1·44	−1·17
1807	−0·21	−0·08	1837	−1·62	−1·53
1810	+0·56	+0·61	1840	+1·73	+1·31

"The greatest difference in the above table, viz. that for 1771, is deduced from a single observation, whereas the difference immediately preceding, which is deduced from the mean of several observations, is much smaller. The error of the tables for 1780 is found by interpolating between the errors given by the observations of 1781, 1782, and 1783, and those of 1769 and 1771. The differences between the results of the two hypotheses are exceedingly small till we come to the last years of the series, and become sensible precisely at the point where both sets of results begin to diverge from the observations; the errors corresponding to the second hypothesis being, however, uniformly smaller. The errors given by the *Greenwich Observations* of 1843 are very sensible, being for the first hypothesis + 6″·84, and for the second + 5″·50. By comparing these errors, it may be inferred that the agreement of theory and observation, would be rendered very close by assuming $\frac{a}{a^1} = 0{\cdot}57$, and the corres-

ponding mean longitude on the 1st October, 1846, would be about 315° 20', which I am inclined to think is not far from the truth. It is plain also that the eccentricity corresponding to this value of $\frac{a}{a^1}$, would be very small. In consequence of the divergence of the results of the two hypotheses, still later observations would be most valuable for correcting the distances, and I should feel exceedingly obliged if you would kindly communicate to me two normal places near the oppositions of 1844 and 1845.

"As Flamsteed's first observation of *Uranus* (in 1690) is a single one, and the interval between it and the rest is so large, I thought it unsafe to employ this observation in forming the equations of condition. On comparing it with the theory, I find the difference to be rather large, and greater for the second hypothesis than for the first, the errors being $+ 44''{\cdot}5$ and $+ 50''{\cdot}0$ respectively. If the error be supposed to change in proportion to the change of mean distance, its value corresponding to $\frac{a}{a^1} = 0{\cdot}57$, will be about $+ 70''$, and the error in the time of transit will be between 4^s and 5^s. It would be desirable to ascertain whether Flamsteed's manuscripts throw any light on this point.

"The corrections of the tabular radius vector of *Uranus*, given by the theory for some late years, are as follows:—

Date	Hypoth. I.	Hypoth. II.
1834	+ 0·005051	+ 0·004923
1840	+ 0·007219	+ 0·006962
1846	+ 0·008676	+ 0·008250

"The correction for 1834 is very nearly the same as that which you have deduced from observation, in the *Astronomische Nachrichten;* but the increase in later years is more rapid than the observations appear to give it: the second hypothesis, however, still having the advantage.

"I am at present employed in discussing the errors in latitude, with the view of obtaining an approximate value of the inclination and position of the node of the new planet's orbit; but the perturbations in latitude are so very small that I am afraid the result will not have great weight. According to a rough calculation made some time since, the inclination appeared to be rather large, and the longitude of the ascending node to be about 300°; but I am now treating the subject much more completely, and hope to obtain the result in a few days.

"I have been thinking of drawing up a brief account of my investigation to present to the British Association."

Mr. Main, acting for the Astronomer Royal in his absence, answered this letter as follows:—

No. 21. *The Rev. R. Main to J. C. Adams, Esq.*

"*Royal Observatory, Greenwich*, 1846, *Sept.* 5.

"The Astronomer Royal is not at home, and he will be absent for some time; but it appears to me of so much importance that you should have immediately the normal errors of *Uranus* for 1844 and 1845, that I herewith send you the former (the volume for 1844 has been published for some time), and I shall probably be able to send you those for 1845 on Tuesday next, as I have given directions to have the computations finished immediately. If a place (geocentric) for the present year should be of value to you, I could probably send one in a few days."

In acknowledging this letter, Mr. Adams used the following expression:—

No. 22.   *J. C. Adams, Esq. to the Rev. R. Main*
[EXTRACT.]

"St. John's College, Cambridge, 7th Sept. 1846.

"I hope by to-morrow to have obtained approximate values of the inclination and longitude of the node."

On the same day, Sept. 7, Mr. Main transmitted to Mr. Adams the normal places for 1845, to which allusion was made in the letter of Sept. 5.

On the 31st of August, M. Le Verrier's second paper on the place of the disturbing planet (the third paper on the motion of *Uranus*) was communicated to the French Academy. I place the notice of this paper after those of September 2, &c. because, in the usual course of transmission to this country, the No. of the *Comptes Rendus* containing this paper would not arrive here, at the earliest, before the third or fourth week in September; and it does not appear that any earlier notice of its contents was received in England.

It is not my design here to give a complete analysis of this remarkable paper; but I may advert to some of its principal points. M. Le Verrier states that, considering the extreme difficulty at attempting to solve the problem in all its generality, and considering that the mean distance and the epoch of the disturbing planet were determined approximately by his former investigations, he adopted the corrections to these elements as two of the unknown quantities to be investigated. Besides these, there are the planet's mass, and two quantities from which the excentricity and the longitude of perihelion may be inferred; making, in all, five unknown quantities depending solely on the orbit and mass of the disturbing planet. Then there are the possible corrections to the mean distance of *Uranus*, to its epoch of longitude, to its longitude of perihelion, and to its acentricity; making, in all, nine unknown quantities. To obtain these, M. Le Verrier groups all the observations into thirty-three equations. He then explains the peculiar method by which he derives the values of the unknown quantities from these equations. The elements obtained are,—

Semi-axis Major	36·154	$\left(\text{or } \dfrac{a}{a^1} = 0\text{·}531. \right)$
Periodic Time	217^y·387	
Excentricity	0·10761	
Longitude of Perihelion....................	284° 45'	
Mean Longitude, 1 Jan. 1847.............	318 47	

$$\text{Mass } \frac{1}{9300} = 0\text{·}0001075$$

True Heliocentric Longitude 1 Jan. 1847	326° 32'
Distance from the Sun..........................	33·06

It is interesting to compare these elements with those obtained by Mr. Adams. The difference between each of these and the corresponding element obtained by Mr. Adams in his second hypothesis is, in every instance, of that kind which corresponds to the further change in the assumed mean distance recommended by

Mr. Adams. The agreement with observations does not appear to be better than that obtained from Mr. Adams's elements, with the exception of Flamsteed's first observation of 1690, for which (contrary to Mr. Adams's expectation) the discordance is considerably diminished.

M. Le Verrier then enters into a most ingenious computation of the limits between which the planet may be sought. The principle is this: assuming a time of revolution, all the other unknown quantities may be varied in such a manner, that though the observations will not be so well represented as before, yet the errors of observation will be tolerable. At last, on continuing the variation of elements, one error of observation will be intolerably great. Then, by varying the elements in another way, we may at length make another error of observation intolerably great; and so on. If we compute, for all these varieties of elements, the place of the planet for 1847, its *locus* will evidently be a discontinuous curve or curvilinear polygon. If we do the same thing with different periodic times, we shall get different polygons; and the extreme periodic times that can be allowed will be indicated by the polygons becoming points. These extreme periodic times are 207 and 233 years. If we now draw one grand curve, circumscribing all the polygons, it is certain that the planet must be within the curve. In one direction, M. Le Verrier found no difficulty in assigning a limit; in the other he was obliged to restrict it, by assuming a limit to the excentricity. Thus he found that the longitude of the planet was certainly not less than 321°, and not greater than 335° or 345°, according as we limit the excentricity to 0·125 or 0·2. And if we adopt 0·125 as the limit, then the mass will be included between the limits 0·00007 and 0·00021; either of which exceeds that of *Uranus*. From this circumstance, combined with a probable hypothesis as to the density, M. Le Verrier concluded that the planet would have a visible disk, and sufficient light to make it conspicuous in ordinary telescopes.

M. Le Verrier then remarks, as one of the strong proofs of the correctness of the general theory, that the error of radius vector is explained as accurately as the error of longitude. And finally, he gives his opinion that the latitude of the disturbing planet must be small.

My analysis of this paper has necessarily been exceedingly imperfect, as regards the astronomical and mathematical parts of it; but I am sensible that, in regard to another part, it fails totally. I cannot attempt to convey to you the impression which was made on me by the author's undoubting confidence in the general truth of his theory, by the calmness and clearness with which he limited the field of observation, and by the firmness with which he proclaimed to observing astronomers, "Look in the place which I have indicated, and you will see the planet well." Since Copernicus† declared that, when means should be discovered for improving the vision, it would be found that *Venus* had phases like the Moon, nothing (in my opinion) so bold, and so justifiably bold, has been uttered in astronomical prediction. It is here, if I mistake not, that we see a character far superior to that of the able, or enterprising, or industrious mathematician; it is here that we see the philosopher. The mathematical investigations will doubtless be published in detail; and they will, as mathematical studies, be highly instructive: but no details published after the planet's discovery can ever have for me the charm which I have found in this abstract which preceded the discovery.

---

† I borrow this history from Smith's *Optics*, sect. 1050. Since reading this Memoir, I have, however, been informed by Professor De Morgan, that the printed works of Copernicus do not at all support this history, and that Copernicus appears to have believed that the planets are self-luminous.—G.B.A.

I understand that M. Le Verrier communicated his principal conclusions to the astronomers of the Berlin Observatory on September 23, and that, guided by them, and comparing their observations with a star-map, they found the planet on the same evening. And I am warranted by the verbal assurances of Professor Challis in stating that, having received the paper on September 29, he was so much impressed that the sagacity and clearness of M. Le Verrier's limitations of the field of observation, that he instantly changed his plan of observing, and noted the planet, as an object having a visible disk, on the evening of the same day.

My account, as a documentary history, supported by letters written during the events is properly terminated; but I think it advisable, for the sake of clearness, to annex extracts from a letter which I have received from Professor Challis since the beginning of October, when I returned to England.

No. 23.   *Professor Challis to G. B. Airy.*

[EXTRACT.]

"Cambridge Observatory, October 12, 1846.

"I had heard of the discovery [of the new planet] on October 1. * * * I find that my observations would have shewn me the planet in the early part of August, if I had only discussed them. I commenced observing on July 29, attacking first of all, as it was prudent to do, the position which Mr. Adams' calculations assigned as the most probable place of the planet. On July 30, I adopted the method of observing which I spoke of to you. * * * In this way I took all the stars to the 11th magnitude in a zone of 9' in breadth, and was sure that none brighter than the 11th escaped me. My next observations were on August 4. On this day * * * I took stars here and there in a zone of abut 70' in breadth, purposely selecting the brighter, as I intended to make them reference-points for the observations in zones of 9' breadth. Among these stars was the planet. A comparison of this day's observations with a good star-map would almost probably have detected it. On account of moonlight I did not observe again till August 12. On that day I went over again the zone of 9' breadth which I examined on July 30. * * * The space gone over on August 12, exceeded in length that of July 30, but included the whole of it. On comparing [at a later time] the observations of these two days, I found that the zone of July 30 contained *every* star in the corresponding portion of the zone of August 12, *except one star of the 8th magnitude.* This, according to the principle of search, which in the want of a good star-map I had adopted, must have been a planet. It had wandered into the latter zone in the interval between July 30 and August 12. By this statement you will see, that, after four days of observing, the planet was in my grasp, if only I had examined or mapped the observations. I delayed doing this, partly because I thought the probability of discovery was small till a much larger portion of the heavens was scrutinised, but chiefly because I was making a grand effort to reduce the vast number of comet observations which I have accumulated; and this occupied the whole of my time when I was not engaged in observing. I actually compared to a certain extent the observations of July 30 and August 12, soon after taking them, more for the sake of testing the two methods of observing adopted on those days than for any other purpose; and I stopped short within a very few stars of the planet. After August 12, I continued my observations with great diligence, recording the positions of, I believe, some thousands of stars: but I did not again fall in with the planet, as I took positions too early in right ascension. * * * On Sept. 29, however, I saw, for the first time, Le Verrier's last results, and on the evening of that day I observed strictly according to his suggestions, and within the limits he recommended; and I was also on the lookout for a disk. Among 300 stars which I took that night, I singled out one, against which I directed my assistant to note "seems to have a disk," which proved to be the planet. I used on this, as on all other occasions, a power of 160. This was the third time I obtained an approximate place of the planet before I heard of its discovery."

This letter was written to me purely as a private communication, but I have received permission from Professor Challis to publish it with the rest.

Before terminating this account, I beg leave to present the following remarks:—

First. It would not be just to institute a comparison between papers which at this time exist only in manuscript, and papers which have been printed by their authors; the latter being in all cases more complete and more elaborately worked out than the former.

Second. I trust that I am amply supported, by the documentary history which I have produced, in the view which I first took, namely, that the discovery of this new planet is the effect of a movement of the age. It is shewn, not merely by the circumstance that different mathematicians have simultaneously but independently been carrying on the same investigations, and that different astronomers, acting without concert, have at the same time been looking for the planet in the same part of the heavens; but also by the circumstance that the minds of these philosophers, and of the persons about them, had long been influenced by the knowledge of what had been done by others, and of what had yet been left untried; and that in all parts of the work the mathematician and the astronomer were supported by the exhortations and the sympathy of those whose opinions they valued most. I do not consider this as detracting in the smallest degree from the merits of the persons who have been actually engaged in these investigations.

Third. This history presents a remarkable instance of the importance, in doubtful cases, of using any received theory as far as it will go, even if that theory can claim no higher merit than that of being plausible. If the mathematicians whose labours I have described had not adopted Bode's law of distances (a law for which no physical theory of the rudest kind has ever been suggested), they would never have arrived at the elements of the orbit. At the same time, this assumption of the law is only an aid to calculation, and does not at all compel the computer to confine himself perpetually to the condition assigned by this law, as will have been remarked in the ultimate change of mean distance made by both the mathematicians, who have used Bode's law to give the first approximation to mean distance.

Fourth. The history of this discovery shews that, in certain cases, it is advantageous for the progress of science that the publication of theories, when so far matured as to leave no doubt of their general accuracy, should not be delayed till they are worked to the highest imaginable perfection. It appears to be quite within probability, that a publication of the elements obtained in October 1845 might have led to discovery of the planet in November 1845.

I have now only to request the indulgence of my hearers for the apparently egoistical character of the account which I have here given; a character which it is extremely difficult to remove from a history that is almost strictly confined to transactions with which I have myself been concerned.

References and Bibliography

CHAPTER 1

Burgess, E., 1988. *Uranus and Neptune: the Distant Giants*. New York.
Greeley, R., 1987. *Planetary Landscapes*. Boston and London.
Moore, P., 1995. *The Guinness Book of Astronomy*. London. (For data, tables etc.)
Moore, P., and Hunt, G. E., 1988. *Atlas of Uranus*. Cambridge.
Moore, P., and Hunt, G. E., 1994. *Atlas of Neptune*. Cambridge.

CHAPTER 2

Adams, J. C., 1841. In manuscripts dealing with the perturbations of Uranus, 1841–46. In the Library of St. John's College, Cambridge University. See also *Occ. Notes R. A. S.* **2** No. 11, 47.
Adams, J. C., 1844. Letter, dated 18 December 1844, made available to me by Mrs Norma Foster, and in her possession.
Adams, J. C., 1846. An Explanation of the observed Irregularities in the Motion of Uranus, on the Hypothesis of Disturbances caused by a more distant Planet; with a Determination of the Mass, Orbit and Position of the disturbing Body. *MNRAS* **7** 149.
Adams, J. C., 1847. The History of the Discovery of Neptune. *Mem. RAS* **16**.
Adams, J. C., 1876. Presidential Address. *MNRAS* **36** 232.
Adams, W. G., 1896. *The scientific papers of John Couch Adams*, 2 vols. Cambridge.
Airy, G. B., 1846. Account of some circumstances historically connected with the discovery of the Planet exterior to Uranus. *MNRAS* **7** 121–144.
Airy, G. B., 1847. Mr. Adams and the New Planet. (Letter). *Athenæum,* 20 February 1847, p. 199.
Airy, G. B., 1847. Name of the new planet. *Athenæum,* 27 February 1847, p. 229.
Airy, W. (editor), 1896. *Autobiography of Sir George Biddell Airy*.
Alexander, A. F. O'D., 1965. *The Planet Uranus*. London.
Allen, E., 1892. Letter about the discovery of Neptune. *Observatory* **15** 260.
Arago, F., 1846a. Examen des remarques critiques et des questions de priorité que la découverte de M. Le Verrier a soulevées. *Comptes rendus* **23** 751, Paris.

Arago, F., 1846b. Letter about the name of Neptune. *Astr. Nach.* **25** 81.

Arago, F., 1847. (Comments on name of Neptune) *Astr. Nach.* **25** 159.

Bessel, F. W., 1840. *Populäre Vorleschungen,* p. 448.

Biot, J.-B., 1847. Sur la planète, nouvellement découverte par M. L.e Verrier, comme consequence de la théorie d'attraction. *Jnl. des Savants* **12** series 3, 64–85, Paris.

Bouvard, A., 1821. Tables astronomiques publiées par le Bureau des Longitudes de France contenant les Tables de Jupiter, de Saturne et d'Uranus construites d'après la théorie de la Mécanique celeste, p.vix.

Brookes, C., 1970. On the prediction of Neptune. *Celest. Mech.* **3** 67–80

Challis, J., 1846a. Account of observations at the Cambridge Observatory for detecting the Planet exterior to Uranus. *MNRAS* **7** 145.

Challis, J., 1846b. Letter to Arago. *Comptes rendus* **23** 715.

Challis, J., 1846c. Report to the Cambridge Observatory, 12 December, p. 2.

Chapman, A., 1988. Private Research and Public Duty: George Biddell Airy and the search for Neptune. *Jnl. Hist. Astronomy* **19** 121 ff.

Danjon, A., 1946. La découverte de Neptune. *Bull. Soc. Astr. de France* **60** 255 ff.

De Broglie, L., 1946. La découverte de Neptune et la science moderne. *Bul. Soc. Astr. de France* **60** 246 ff.

Dewhirst, D. W., 1976. *Vistas in Astronomy* **20** 109.

Dewhirst, D. W., 1985. *Vistas in Astronomy* **28** 147.

Dreyer, J. L. E., 1882. Historical note concerning the discovery of Neptune. *Copernicus* **2** 63.

Ellis, W., 1905. *Observatory* **28** 181.

Encke, J., 1846a. Account of the Discovery of the Planet of Le Verrier at Berlin. *MNRAS* **7** 153.

Encke, J., 1846b. Manuscript letter in the Paris library.

Encke, J., 1846c. Letter to the Editor, *Astr. Nach.* No 580.

Encke, J., 1848. *Comptes rendus* **23** 602.

Galle, J., 1846. Manuscript letter in the Paris library.

Galle, J., 1882. Ueber die erste Auffindung des Planeten Neptun. *Copernicus* **2** 96.

Gingerich, O., 1958. The naming of Uranus and Neptune. Leaflet *Astr. Soc. Pacific* **352** 1 ff.

Gould, B. A., 1850. Report on the history of the discovery of Neptune. Annex to 1850 volume of the report of the Board of Regents of the Smithsonian Institution, Washington.

Grant, R. H., 1852. *History of Physical Astronomy,* London, p. 196.

Grosser, M., 1962. *The Discovery of Neptune.* Harvard.

Herschel, J., 1846. Le Verrier's Planet. *Athenæum* 3 October 1846.

Herschel, J., 1848. *MNRAS* **8** 111.

Hind, J. R., 1846. (Letter). *The Times,* London, 1 October, p. 8.

Holden, E. S., 1892. Historical note relating to the search for the planet Neptune in England, 1845–46. *Astron. Astrophys.* **11** 287. See also *Nature* **45** 522 (1892).

Hunt, G. E., and Moore, P., 1988. *Atlas of Uranus.* Cambridge.

Jackson, J., 1949. The discovery of Neptune: a defence of Challis. *Month. Not. Astron. Soc. South Africa* **55** 312.

Jackson, J., 1955. The discoveries of Neptune and Pluto. *Observatory* **19** 364.

Jones, H. Spencer, 1946. *John Couch Adams and the Discovery of Neptune.* Cambridge.

Jones, H. Spencer, 1947. G. B. Airy and the Discovery of Neptune. *Nature* **158** 830.

Lassell, W., 1846. Notebook 9.7, entry for 2 October 1846. Lassell Papers, Royal Astronomical Society.

Le Verrier, U. J. J., 1842. Seconde note sur les perturbations de la planète Uranus. *Comptes rendus* **14** 660.

Le Verrier, U. J. J., 1845. Premier mémoire sur la théories d'Uranus. *Comptes rendus* **21** 105.

Le Verrier, U. J. J., 1846a. Recherches sur les movements d'Uranus. *Comptes rendus* **22** 907.

LeVerrier, U. J. J., 1846b. Sur la planète qui produit les anomalies observées dans le movement d'Uranus — Determination de sa masse, de son orbite et de sa position actuelle. *Comptes rendus* **23** 428, 657.

Le Verrier, U. J. J., 1846c. Comparison des observations de la nouvelle planète avec la théorie déduite des perturbations d'Uranus. *Comptes rendus* **23** 771.

Le Verrier, U. J. J., 1846d. Recherches sur le movement de la planète Herschel (dite Uranus). *Con. des Temps for 1849.* Additions, p. 3–254.

Le Verrier, U. J. J., 1846e. Letter to Galle, written 18 September 1846. *Nature* **85**, 184 (1910). See also See, T. J. J., *Pop. Astr.* October 1910, p. 475.

Lynn, W. T., 1896. The jubilee of the discovery of Neptune. *Observatory* **19** 364.

Lyttleton, R. A., 1958. A short method for the discovery of Neptune. *MNRAS* **118** 551 ff.

Mädler, J. H. von, 1841. *Populäre Astronomie*, p. 345.

Mitchel, O. M., 1846. The new planet: its name and discovery. *Sidereal Messenger* **1** 60.

Moore, P., 1986. *William Herschel: astronomer of Bath.* William Herschel Society, Bath.

Nichol, J. P., 1849. *The Planet Neptune.* Edinburgh.

Nicolai, F. G. B., 1836. *Astr. Nach.* **13** 94.

Nieto, M. M., 1972 *The Titius–Bode Law of Planetary Distances.* Oxford.

Rines, D., 1912. The discovery of the planet Neptune. *Pop. Astr.* **20** 482 ff.

Royal Astronomical Society, 1847. *MNRAS* **7** 216.

Sampson, R. A., 1901. A description of Adams' ms. on the perturbations of Uranus. London.

Schumacher, H. C., 1846. Circular announcing the discovery of Neptune: letter from J. F. Encke. Altona, 29 September 1846.

Sheepshanks, R., 1847. On the Planet Neptune and the Royal Astronomical Society's medal. London.

Smart, W. M., 1946. John Couch Adams and the Discovery of Neptune. *Nature* **158** 648 ff. (Reply by Spencer Jones, **158**, 830).

Smart, W. M., 1947. John Couch Adams and the Discovery of Neptune. *Occasional Notes, Royal Astronomical Society* **2** No 11, 33 ff.

Smith, R. W., 1983. William Lassell and the Discovery of Neptune. *Jnl. for Hist. of Astronomy* **14** 30.

Standish, E. M., 1981. *Nature* **290** 164.

Struve, F. G. W., 1846. Ueber den neuen Hauptplaneten Neptune. St. Petersburg.

Struve, F. G. W., 1849. *Bull. de la classe physico-mathématique de l'Académie des Sciences de St. Petersbourg* **7** cols. 321–336.

Struve, O., 1951. The telescopic discovery of Neptune. *Jnl. Brit. Astron. Assoc.* **61** 166.

Troy, H. S., 1969. *The Adamses of Lidcot.* Wordens of Cornwall, Penzance, p.3.

Turner, H. H., 1963. *Astronomical Discovery.* California.

Valz, J. E. B., 1835. Letter on Halley's Comet. *Comptes rendus* **1** 130.

Vincent, M., 1958. Le Mystère de la découverte de Neptune par Le Verrier. Fischbacher, Paris.

Box in the Cambridge University Library, containing the Adams papers. This list was kindly made available to me by Dr. D. W. Dewhirst.

Papers relating to the discovery of Neptune 1845–47 are contained in this box. There are in addition a number of letters of minor significance, containing passing references to Neptune, scattered through the ordinary archive letter-files for these years.

D. W. Dewhirst
1967 January

I. *Letters ante-dating the discovery*

1. Airy to Challis 1846 June 30.
   As touching the recommendations . . .
   Letter on 1 sheet, folded.
2. Airy to Challis 1846 July 9. From Ely.
   You know that I attach importance . . .
   Letter on 1 sheet, folded.
3. Airy to Challis 1846 July 13.
   I have drawn up the inclosed paper . . .
   Letter on 1 sheet, folded.
4. Airy ms. of July 12.
   Suggestions for the Examination of a position of the Heavens . . .
   (Proposals for search, 4 sides on 1 folded sheet.)
5. Airy to Challis 1846 July 21.
   I am very glad . . .
   Letter on 1 sheet, folded.
6. Airy to Challis 1846 Aug. 6.
   I do not know when you may desire . . .
   Letter on 1 sheet, folded.
7. Main to Challis 1846 Aug. 8.
   I have referred your note to the A.R. . . . .
   Letter on 1 sheet.
8. James Breen to Challis 1846 Aug. 8.
   I had the honor to receive . . .
   Letter on 1 sheet.
9. Main to Challis 1846 Aug. 15.
   I have great pleasure in acceding . . .

Letter on 1 sheet.
10. J. R. Hind to Challis 1846 Sept. 16.
   I have received a letter this morning from M. Faye . . .
   Letter of 4 sides on 2 sheets including cover and seal.
11. Airy to Challis 1846 Aug. 25. From Weisbaden.
   Before I left England . . .
   Letter on 1 sheet with Airy's seal.

II. *Letters after the discovery*

12. J. R. Hind to Challis 1846 Sept. 30 13^h.
   Le Verrier's planet is discovered. Dr. Galle . . .
   Letter on 1 sheet, folded.
13. Airy to Challis 1846 Sept. 30. From Gotha.
   While I was sitting yesterday at dinner . . .
   Letter on 1 sheet, folded.
14. Main to Challis 1846 Oct. 1.
   In the absence of the A . . . R . . .
   Letter on 1 sheet.
15. Challis 1846 Oct. 1.
   Discovery of a new planet . . .
   Letter in *Cambridge Chronicle,* letter dated Oct. 1. (Press cutting.)
16. Challis 1846 Oct. 16.
   The New Planet . . .
   Letter in *Cambridge Chronicle,* letter dated Oct. 16. (Press cutting.)
17. Nichol to Challis 1846 Oct. 7.
   I have indulged the hope of seeing you at Cambridge . . .
   Letter on 1 sheet, folded.
18. Airy to Challis.
   I returned from Hamburgh . . .
   Letter on 1 sheet, folded.
19. C. Piazzi Smyth to Challis 1846 Oct. 20.
   I have to thank you, I believe . . .
   Letter on 1 sheet.
20. Airy to Challis 1846 Oct. 30.
   I am quite ashamed of having given you . . .
   Letter on 1 sheet, folded.
21. Airy to Challis 1846 Nov. 5.
   I received today the Guardian . . .
   Letter on 1 sheet, folded.
22. Schumacher to Challis 1846 Nov. 10. From Altona.
   I have duly received the two letters you favoured me with . . .
   Letter on 1 sheet, folded.
23. Airy to Challis 1846 Nov. 19.
   I hand you note [sic] (about copies of my homily . . .)
   Letter on 1 sheet.
24. Sheepshanks to Challis. (1846?) Nov. 20. From Reading.

You shall have as many copies for Cambridge as you like . . .
Letter on 1 sheet.

25. W. S. Stratford to Challis 1846 Dec. 16.
I have this day forwarded to your Observatory . . .
Letter on 1 sheet.

26. Sheepshanks to Challis 1846 Dec. 18.
Thank your for your statement . . .
Letter on 1 sheet.

26a. Airy to Challis 1846 Dec. 18.
The alarming illness of one of my children . . .

26b. I am much obliged by the copy of your report . . .
Two separate letters on 1 folded sheet.

27. Airy to Challis 1846 Dec. 21.
Our little girl Hilda continues very ill . . .
Letter on 1 sheet.

28. Edward J. Cooper to Challis 1846 Dec. 30.
Your interesting paper upon the proceedings . . .
Letter on 1 sheet.

29. J. Glaisher to Challis 1847 (presumably, tho' dated 1846) Jan. 4.
I beg to thank you for the copy . . .
Letter on 1 sheet.

30. Challis ms. 1847 Feb. 1.
I felt very much obliged by the note you addressed . . .
Draft letter on 1 sheet ? of his reply to Schumacher, item 22 above.

III. *Other letters and mss. by Adams and others*

31. Enveloped addressed "The Revd. Profr. Challis . . . With Mr. Adams'
Compts." containing torn off slip of paper "M. Verrier recommends exploring
for the new Planet the region of the Ecliptic from 321° to 335° of heliocentric
Longitude". (This slip and envelope have been associated since I first saw the
Neptune Papers — 1951, but the hands are not the same and I do not think the
slip is Adams' handwriting? — D.W.D.)

32. Undated letter from Adams (to Challis?).
The Elements of the New Planet I make to [sic] as follows . . .
1 sheet folded, bears at top in Challis's hand "(Received in Sept. 1845)".

33. Undated letter Sheepshanks to Challis (in pencil at top 1846?).
You need give yourself little trouble . . .
Letter on 1 sheet folded.

34. Unidentified scrap of paper starting "No. 648 de l'Institut" . . .

35. 3 sheets written on one side, in Adams' hand.
If we suppose No. 1007 of the British Catalogue . . .
Hel. Long. of the Planet Augt. 8·0 . . .
Mean Dist. = 30·35 with a p . . . e . . .
The first of these, signed by J.C.A., is the "list of predicted places Mr. Adams
drew up for me" (Challis, Cambridge Observations XV Appendix 1, p. 2).

36. 4 sheets, in Challis's hand.
Drafts for various pages of Cambridge Observations XV Appendix 1.

37. 17 sheets, in Challis's hand.
A draft version of Challis's "Account" (Mem. RAS **16**).
38. Printed Circular from Schumacher, Altona 1846 Sept. 29.
Circular. Schreiben des Herrn Professors Encke . . .
1 sheet, folded.

IV. *Notebooks, etc.*

39. Notebook.
Zones observed with Northumberland Telescope. Search for Neptune.
(See esp. p. 7 for Aug. 12 "This is the planet"; and the fourth written page back
from the end "seemed to have a disc".)
40. Notebook.
Observations of Neptune & etc.
(Entirely calculations & reductions. See esp. for Aug. 4, p. 8.)
41. 3 forms of calculation of an ephemeris, and a ms. star map.
Probably Challis's hand: nature unidentified.

V. *Addendum.* Some months after the above material had been collated a number
of additional items came to light and are here appended.

42. Airy to Challis 1845 Sept. 29.
I was, I suppose, on my way from France when Mr. Adams called . . .
Letter on 1 sheet.
43. Wm. Lassell to Challis 1847 Jan. 19.
In reply to your note of the 16th . . . (suspected ring about Neptune)
Letter on 1 sheet folded.
44. W. R. Dawes to Challis 1847 April 7.
I have to thank you for a copy of your 'Second Report . . .
Letter on 1 sheet folded.
45. W. Struve to Challis 1847 Jan. 23/Feb. 4.
I take the liberty of sending you a little note . . .
Letter on 1 sheet folded.
46. Sur la denomination de la planète nouvellement decouverte . . .
The note referred to in 45; printed pamphlet on 1 sheet folded, Poulkova
17/29 Dec. 1846.
47. Challis to Struve 1847 March 10.
Accept my best thanks for the copy of your communication . . .
Draft copy of a reply to 46, on two scraps of paper.

CHAPTER 3

Albers, S. C., 1979. *Sky and Tel.* **57** 220.
Bond, W. C., 1847. Letter to the Editor, *Astr. Nach.* No. 591; and Observations of
Neptune made at the Cambridge Observatory, *Astr. Nach.* No. 595.
Brookes, C. J., 1972. On a Criterion for the Prediction of Neptune. *MNRAS* **158** 79.
Challis, J., and Thompson, R. A., 1848. Observations of the positions of Neptune.
MNRAS **8** 6.

Eklöf, O., 1946. Neptunus upptäckt — ett hundraårsimme. *Populär Astronomik Tidsskrift* **27** 375 ff.

Encke, J., 1847. *Astr. Nach.* No. 588.

Gould, B. A., 1850. *Report on the History of the Discovery of Neptune.* Smithsonian Institution, Washington.

Gyldén, J., 1861. Beräkning af en teori for planeten Neptunus. Helsingfors.

Hind, J. R., 1850. Unnoticed observations of Neptune. *MNRAS* **10** 42.

Hind, J. R., 1851. Letter on Lamont's third observation of Neptune. *MNRAS* **11** 11.

Hough, G. W., 1863. Observations with the Olcott meridian circle at Dudley Observatory, Albany. *MNRAS* **23** 22.

Hughes, D., 1980. Galileo saw Neptune in 1612. *Nature* **287** 277.

Kowal, C. T., and Drake, S., 1980. Galileo's observations of Neptune. *Nature* **287** 311 ff.

Kowalski, M., 1856. Recherches sur les mouvements de Neptune. Sborniki outschenikki statet déjà cité. *Kazani* **1** 97.

Lalande, J. J., 1801. *Histoire céleste française.* Paris, p. 158.

Lamont, J., 1850. *MNRAS* **10** 42.

Lamont, J., 1851. *MNRAS* **11** 11.

Liais, E., 1866. History of the discovery of the planet Neptune. From: Tischner, A. *L'espace céleste et la natur tropical.* Paris 1866; publ. Gustav Fock, Leipzig 1892. (In French. The quotes given here are from my translation. — P.M.)

Lynn, W. T., 1895. Lamont's observations of Neptune. *Observatory* **18** 27.

Mädler, J. H. von, 1870. Die Entdeckung des Neptun. *Reden und Abhandlungen.* Mädler, Berlin.

Mauvais, F. V., 1847. *Astr. Nach.,* No. 607.

Newcomb, S., 1865. On Kowalski's theory of Neptune. *MNRAS* **25** 45 ff.

Newcomb, S., 1867. An investigation into the orbit of Neptune. *Smithsonian Contrib. to Knowledge* **15** No. 2, Washington.

Petersen, A. C., 1847. Nachsuchtung fruherer Beobtung des Le Verreirschen Planeten. *Astr. Nach.* Nos 594, 595; and Everett, 1847. On the New Planet Neptune, *Astr. Nach.* No. 599. See also Walker, S. C., 1847. Letter to J. Henry, *Astr. Nach.* No. 605.

Peirce, B., 1847a. Investigation into the action of Neptune on Uranus. *Proc. Amer. Acad. Arts and Sciences* **1** 65.

Peirce, B., 1847b. Formulae in the theory of Neptune. *Proc. Amer. Acad. Arts and Sciences* **1** 278 ff.

Peirce, B., 1848a. *MNRAS* **8** 36, 202.

Peirce, B. 1848b. Perturbations of Uranus by Neptune. *Proc. Amer. Philosoph. Soc.* **5** 15.

Rawlins, D., 1970. The great unexplained residual in the orbit of Neptune. *Astron. J.* **75** 856.

Rawlins, D., 1981. Galileo's observations of Neptune. *Nature* **290** 164.

Sampson, R. A., 1901. A description of Adams' manuscripts on the perturbations of Uranus.

Valz, J. E. B., 1847., Éléments provisoires de la planète de M. Le Verrier pour l'époque du 7 décembre 1846. *Comptes rendus* **24** 33, Paris.

Walker, S. C., 1847a. On the new planet Neptune. In: Everett, *Astr. Nach.* **25** 381.

Walker, S. C., 1847b. Elements of the new planet Neptune. *Proc. Amer. Philosoph. Soc.* **4** 332.

Walker, S. C., 1847c. Elliptic elements of the planet Neptune. *Proc. Amer. Philosoph. Soc.* **4** 378.

Walker, S. C., 1848. Investigations which led to the detection of the coincidence between the computed place of the planet Le Verrier and the observed place of a star recorded by Lalande in May 1795. *Trans. Amer. Philosoph. Soc.*, n.s., **10** 141 ff.

CHAPTER 4

Baum, R. M., 1973. *The Planets: Some Myths and Realities.* Newton Abbot, pp. 120–146.

Bond, W. C., 1848. *MNRAS* **8** 9.

Challis, J., 1847a. Second report of the proceedings in the Cambridge Observatory, relating to the new planet (Neptune). *Astr. Nach.* **25** cols 309–314, col. 310.

Challis, J., 1847b. *Astr. Nach.* **25** 231.

Herschel, J., 1846. Letter to Lassell, 1 October 1846. Herschel Papers, Royal Society, H. S. 22.285.

Hetherington, N. S., 1979. Neptune's supposed ring. *Jnl. Brit. Astron. Assoc.* **90** 20 ff.

Hind, J. R., 1847. *MNRAS* **7** 169; also *Astr. Nach.* **25** 207.

Lassell, W., 1846a. *MNRAS* **7**, 167 ff.

Lassell, W., 1846b. Letter to the *Times,* 14 October 1846.

Lassell, W., 1846c. *MNRAS* **7** 157 and 167.

Lassell, W. 1847. Letter to Challis, 19 January 1847. Cambridge Observatory archives.

Lassell, W., 1852a. *MNRAS* **12** 155.

Lassell, W., 1852b. *MNRAS* **13** 38.

Lassell, W., 1852c. Lassell's notebook, 15 December 1852. Papers 16.4, RASA.

Lassell, W., 1853. *MNRAS* **13** 36 ff.

Royal Astronomical Society, 1847. *MNRAS* **7** 217.

Smith, R. W., and Baum, R. M., 1984. William Lassell and the Ring of Neptune; a case study in instrumental failure. *Jnl. for Hist. of Astronomy* **15** 1 ff.

CHAPTER 5

1. Diameter, form and mass

Adams, J. C., 1849. Note on the mass of Uranus. *MNRAS* **9** 159, 203.

Barnard, E. E., 1902. *Astr. Nach.* **157** No. 3760.

Bixby, J. E., and van Flandern, T., 1969. *Astron. J.* **74** 1220.

Bond, G. P., 1851. *Astr. Nach.* **31** 38.

Bonev, V., 1972. On the oblateness of Neptune. *Physics of the Moon and Planets.* (In Russian.)

Camichel, H., 1953. *Annales d'Astrophysique* **16** 41.

Chambers, G. F., 1889. *Handbook of Astronomy.* p. 652.

Cook, A. H., 1972. The masses, densities and moments of inertia of Uranus and Neptune. *Observatory* **92** 84.

Cook, A. H., 1979. A note on the flattening of Uranus and Neptune. *MNRAS* **187** 39P–43P.

Davies, M. E., Abalakin, V. K., Burša, A., Hunt, G. E., Lieske, J. H., Morando, B., Rapp, R. H., Seidelmann, P. K., Sinclair, A. T., and Tjuflin, Y. S., 1988. Report of the IAU/IAG/COSPAR Working Group on Cartographic Coordinates and Rotational Elements of the Planets and Satellites. (Presented at the IAU General Assembly, Baltimore, August 1988.)

De Marcus, W., 1977. *Icarus* **30** 441.

Denison, E. B., 1866. *Astronomy without Mathematics.* p. 125.

Dick, T., 1852. *The Solar System.* p. 74.

Dollfus, A., 1970. New Optical Measurements of the Diameters of Jupiter, Saturn, Uranus and Neptune. *Icarus* **12** 101.

Freeman, K. C., and Lyngå, G., 1970. *Astrophys. J.* **160** 767; and *Proc. Astron. Soc. Australia* **1** 203–4 (1969).

French, R. G., 1984. *Uranus and Neptune.* NASA CP-2330, 349.

French, R. G., Melroy, P. A., Baron, R. L., Dunham, E. W., Meech, K. J., Mink, D. J., Elliot, J. L., Allen, D. A., Ashley, M. C. B., Freeman, K. G., Erickson, E. F., Goguen, L., and Hammel, H. B., 1985. The 1983 June 15 occultations by Neptune. II. The oblateness of Neptune. *Astron. J.* **90** 2624 ff.

Hatanaka, Y., 1971. Diameter and Flattening of Neptune. *Publ. Astr. Soc. Japan,* **23** 263.

Hind, J. R., 1855. *MNRAS* **15** 47.

Holden, E., and Hall, A., 1876. *Astr. Nach.* **78** 186.

Hughes, D. W., 1979. Flattening of Uranus and Neptune. *Nature* **279** 582.

International Astronomical Union, 1977. Report 16B, p. 60.

Kaiser, F., 1872. *Astr. St. Leiden* **3** 273.

Kovalevsky, J., and Link, F., 1969. *Astron. Astrophys.* **2** 398. (In French.)

Kuiper, G. P., 1949. The diameter of Neptune. *Astrophys. J.* **110** 93.

Lassell, W., 1853. *Astr. Nach.* **36** 97.

Lassell, W., and Marth, A., 1867. *MNRAS* **36** 37.

Laugier, B., 1846. *ABL* **57** (1865), Paris.

Levin, B., 1969. Mass and density of Neptune. *Priorda* No. 6, 77–78.

Mädler, J. H. von, 1847. *Astr. Nach.* **25** 107, 232.

Main, R., 1846. Greenwich Observatory Report for 1846. p. 119.

Moore, J. H., and Menzel, D. L., 1928. *Publ. Astr. Soc. Pacific* **40** 234.

Newcomb, S., 1874. *Smithsonian Contr. to Knowledge* **19** 173.

Newcomb, S., 1875. Washington Observatory Report for 1873. Part 1, p. 63.

Peirce, B., 1848a. *Astr. Nach.* **27** 203.

Peirce, B., 1848b. *MNRAS* **8** 128.

Safford, J., 1862. *MNRAS* **22** 143.

Sharonov, V. V., 1964. *The Nature of the Planets.* Israel Programme for Scientific Translations, p. 173.

Struve, A., 1850. Dorpat, *Beo.* **13**, Anh. 24, 1856.

Struve, O., 1847. *Comptes rendus* **25** 813, Paris.

Taylor, G. E., 1968. New Determination of the Diameter of Neptune. *Nature* **219** 474.

Taylor, G. E., 1969. Neptune (Presidential Address). *Jnl. Brit. Astron. Assoc.* **80** 13 ff.

Taylor, G. E., 1970. *MNRAS* **147** 22 ff.

van Biesbroeck, G., 1957. The mass of Neptune from a new orbit of its second satellite, Nereid. *Astron. J.* **62** 272.

Wallace, L., 1975. *Astrophys. J.* **197** 257.

Young, C. A., 1891. *General Astronomy* p. 372.

2. Rotation period

Belton, M. J. S., 1980. *Icarus* **42** 71.

Belton, M. J. S., Wallace, L., Hayes, S. H., and Price, M. J., 1980. Neptune's Rotation Period; a correction and a speculation on the difference between photometric and spectroscopic results. *Icarus* **42** 71 ff.

Belton, M. J. S., and Terrile, R., 1984. *Uranus and Neptune.* NASA CP-2330, 342.

Brown, R. H., Cruikshank, D., and Tokunaga, A. T., 1980. The Rotation Period of Neptune. *Bull. Amer. Astron. Soc.* **12** 704.

Cruikshank, D., 1977. Rotation of Neptune. *Bull. Amer. Astron. Soc.* **9** 511.

Cruikshank, D., 1978. *Astrophys. J. Lett.* **220** L49–L52.

Drobyshevskii, E. M., 1979. On the Equatorial Flows in Uranus and Neptune. *Soviet Astronomy* **23** 598.

Flammarion, C., 1872. *Études et lectures sur l'astronomie.* Paris, part 3, p. 17 (in French).

Hall, M., 1884a *MNRAS* **44** 257.

Hall, M. 1884b *Nature* **31** 193.

Harris, A. W., 1977. An analytical theory of planetary rotation rates. *Icarus* **31** 168 ff.

Harris, A. W., 1980. Where is Neptune's pole? *Bull. Amer. Astron. Soc.* **12** 70.

Hayes, S. H., and Belton, M. J. S., 1977a. The rotational periods of Uranus and Neptune. *Icarus* **32** 383 ff.

Hayes, S. H., and Belton, M. J. S. 1977b. The rotation periods of Uranus and Neptune. *Bull. Amer. Astron. Soc.* **9** 473.

Howell, R. R., Brown, R. H., Cruikshank, D., Morgan, J. S., and Shaya, E. 1981. Continued studies of the rotation of Neptune. *Bull. Amer. Astron. Soc.* **13** 733.

Hubbard, W. B., 1986. On the oblateness and rotation rate of Neptune's atmosphere. NASA CP-2441, pp. 264–270.

Moore, J. H., and Menzel, D. H. 1928. Preliminary results of spectroscopic observations for the rotation of Neptune. *Publ. Astron. Soc. Pacific* **40** 234.

Moore, J. H., and Menzel, D. H. 1930. *Publ. Astron. Soc. Pacific* **42** 330.

Münch, G., and Hippelein, H., 1980. *Astron. Astrophys.* **81** 189.

Öpik, E., and Liviander, R., 1924. *Nature* **113** 366.

Slavsky, D., Smith, H., Smith, B., and Africano, J., 1977. Rotation period of Neptune. *Bull. Amer. Astron. Soc.* **9** 512.

Slavsky, D., and Smith, H. J. 1978. The rotation period of Neptune. *Astrophys. J. Lett.* **226** L49–L52.

Slavsky, D., and Smith, H. J., 1981. Further evidence for the longer rotation periods of Uranus and Neptune. *Bull. Amer. Astron. Soc.* **13** 733.

Smith, H. J., 1981. Some problems with planetary rotations. *Recent advances in observational astronomy.* 63 ff.

Terrile, R. J., and Smith, B. A., 1983. The rotation rate of Neptune from ground-based CCD imaging. *Bull. Amer. Astron. Soc.* **15** 858.

Wempe, J., 1975. Zur Rotationsperiode des Planeten Neptun. *Sterne* **51** Jahrgang, p. 145. (In German.)

3. Brightness and variability

Aksenov, A. N., and Vdovichenko, V., 1983. The reflectivity of Neptune at λ 0.35–0.67 μm. *Astron. Tsirk.* No. 1269, 5–7. (In Russian.)

Appleby, J. F., 1973. Photometry of Neptune. *Bull. Amer. Astron. Soc.* **5** 29.

Armandorff, T. E., and Radick, R. R., 1982. Photometric variability of Titan, Uranus and Neptune, 1979–82. *Bull. Amer. Astron. Soc.* **14** 865.

Baldwin, J. M., 1908. Photoelectric measurements of Neptune. *MNRAS* **68** 614.

Covault, C. E., and French, L. M., 1985. JHK photometry on Uranus and Neptune candidate occultation stars. *Icarus* **66** 630.

Cruikshank, D. P., 1984. Variability of Neptune. *Uranus and Neptune.* NASA CP-2330, 279 ff.

Cruikshank, D. P., 1985. Variability of Neptune. *Icarus* **64** 107ff.

Hall, M., 1884a. Variations in the light of Neptune, Nov. 29–Dec. 14 1883. *MNRAS* **44** 257.

Hall, M., 1884b. *Nature* **31** 193.

Hall, M., 1885. *Observatory* **8** 26.

Hardorp, J., 1981. The Sun among the Stars. *Astron. Astrophys.* **96** 123.

Lockwood, G. W., 1977. Secular brightness increases of Titan, Uranus and Neptune, 1972–76. *Icarus* **32** 413 ff.

Lockwood, G. W., 1978. Analysis of photometric variations of Uranus and Neptune since 1953. *Icarus* **35** 79 ff.

Lockwood, G. W., and Thompson, D. T., 1986a. Long-term brightness variations of Neptune: a solar cycle effect confirmed. *Bull. Amer. Astron. Soc.* **18** 786.

Lockwood, G. W., and Thompson, D. T., 1986b. Long-term brightness variations of Neptune and the solar cycle modulation of its albedo. *Science* **234** 1543.

Lockwood, G. W., Suess, S. T., and Thompson, D. T., 1980. Correlated variations of planetary albedos and solar-interplanetary parameters. *Solar and interplanetary dynamics.* 163 ff.

Macy, W. W., Sinton, W. M., and Beichman, C. A., 1980. Five-micrometer measurements of Uranus and Neptune. *Icarus* **42** 68.

Muller, G., 1884. *Astr. Nach.* **111** 121; and *Observatory* **7** 264.

Neff, J. S., Ellis, T. A., Apt, J., and Bergstralh, J. T., 1985. Bolometric albedos of Titan, Uranus and Neptune. *Icarus* **62** 425 ff.

Pickering, E. C., 1884. *Observatory* **7** 134.
Pickering, E. C., 1885a. *Observatory* **8** 111.
Pickering, E. C., 1885b. *Nature* **32** 12.

CHAPTER 6

Appleby, J. F., 1985a. Radiative–convective equilibrium models of Uranus and Neptune. *Icarus* **65** 383.

Appleby, J. F., 1985b. Radiative–convective equilibrium models of Uranus and Neptune. NASA CP-2441, p. 252.

Barnard, E. E., 1894. *Astron. J.* Nos. 342, 436, 508.

Belton, M. J. S., and Terrile, R., 1984. Rotational Properties of Uranus and Neptune. *Uranus and Neptune*. NASA CP-2330, p. 238.

Čelebonović, V., 1983. A model of Neptune according to the Savic-Kašanin theory. *Moon and Planets* **29** 185 ff.

Chambers, G. F., 1861. *Handbook of Astronomy*. London, p. 83.

Chambers, G. F., 1890. *Handbook of Astronomy* (4th edn). London, p. 653.

Cragg, T., 1953. In: *Guide to the Planets* (Moore, P. ed.) London and New York, p. 177.

Dermott, S. F., 1979. A comment on the internal structures of Uranus and Neptune. *Bull. Amer. Astron. Soc.* **11** 569.

Fernandez, J. A., and Ip, W. H., 1981. Dynamical evolution of a cometary swarm in the outer planetary region. *Icarus* **47** 1981.

Gordon, R. W., 1983. Craters on Mars and Mercury; a History of Prediction and Observation. In: *Yearbook of Astronomy 1983* (Moore, P. ed.), London and New York.

Greenberg. R., *et al.*, 1984. From icy planetesimals to outer planets and comets. *Icarus* **59** 87.

Gudkova, T. V., 1982. Gravitational field and figure of Uranus and Neptune. *Arhrofiz.i.geokosm.issled.* Moskva, pp. 88–90. (In Russian.)

Gudkova, T. V., 1983. Models, figures and gravitational fields of Uranus and Neptune. Vses. nauchn.stud.konf.po probl. kosmonavtiki; Korolevsk, i, Gagarinsk.chteniya. Tez doki. Ch. 1. p. 52. (In Russian.)

Gudkova, T. V., and Zharkov, V. N., 1984. Models of Uranus and Neptune. *Astron. Vestn.* **18** 293 ff. (In Russian.)

Gudkova, T. V., 1986. Neon-enriched models of Uranus and Neptune. *Prikl.metod.mekh.Mosk. fiz-tech.inst.*Moskva, pp. 49–52. (In Russian.)

Heasley, J. N., Pilcher, C. B., Howell, R. B., and Caldwell, J. J., 1984. Restored methane band images of Uranus and Neptune. *Icarus* **57** 432.

Holmes, M. A. W., 1908. *Nature* **77** 258.

Hubbard, W. B., 1978. Comparative thermal evolution of Uranus and Neptune. *Icarus* **35** 177 ff.

Hubbard, W. B., and MacFarlane, J. J., 1979. Thermal evolution of Uranus and Neptune. *Bull. Amer. Astron. Soc.* **11** 569.

Hubbard, W. B., and MacFarlane, J. J., 1980. Structure and evolution of Uranus and Neptune. *J. Geophys. Res.* **85** 225 ff.

Hunten, D. M., 1984. Atmospheres of Uranus and Neptune. *Uranus and Neptune.* NASA CP-2330, p. 27 ff.

Jeffreys, H., 1923. *MNRAS* **83** 350 ff.

Lewis, J. S., 1984. The origin and evolution of Uranus and Neptune. *Uranus and Neptune.* NASA CP-2330, p. 3 ff.

Makalkin, A. B., 1973. The Structure and Models of Neptune. *Solar System Res.* **6** 153 ff.

Margulis, L., Halvorson, H. O., Lewis, J., and Cameron, A. G. W., 1977. Limitations to the growth of micro-organisms on Uranus, Neptune and Titan. *Icarus* **30** 703 ff.

Murphy, R. E., and Trafton, L. M., 1974. Evidence for an internal heat source in Neptune. *Astrophys. J.* **193** 253.

Neff, J. S., 1985. Bolometric Albedos of Titan, Uranus and Neptune. *Icarus* **62** 425.

Podolak, M., and Cameron, A. G. W., 1974. Models of the Giant Planets. *Icarus* **22** 123.

Podolak, M., and Reynolds, R. T., 1981. On the structure and composition of Uranus and Neptune. *Icarus* **46** 40 ff.

Podolak, M., Young, R., and Reynolds, R. T., 1985. *Icarus* **63** 266.

Pollack, J. B., Podolak, M., Bodenheimer, P., and Christofferson, B., 1987. Planetesimal dissolution in the envelopes of the forming giant planets. *Icarus* **67** 409.

Pollack, J. B., Rages, K., Barnes, K., Bergstralh, J., Wenkert, D., and Danielson, G. E., 1985. Estimates of the bolometric and radiation balance of Uranus and Neptune. *Icarus* **65** 442.

Ramsey, W. H., 1951. *MNRAS* **111** 427 ff.

Ross, M., 1981. *Nature* **292** 435; *Sky and Tel.* **62** 317.

See, T. J. J., 1899. In: Clerke, *History of Astronomy during the Nineteenth Century.* London, p. 206.

Slipher, V. M., 1904. *Nature* **70** 390; and *Lowell Observatory Bulletin,* No. 13.

Smit, B., 1985. Ground-based observations of Uranus and Neptune with PZS equipment. *Astron. Vestn.* **19** 42 ff.

Smith, B. A., Reitsema, H. J., and Larson, S. M., 1979. *Bull. Amer. Astron. Soc.* **11** 570.

Smith, B. A., and Reitsema, H. J., 1982. In: *Uranus and the Outer Planets* (Hunt, G. ed.). Cambridge (England), p. 173.

Smith, B. A., 1984. Near Infrared Imaging of Uranus and Neptune. *Uranus and Neptune.* NASA CP-2330, pp. 216–217.

Stier, M. T., Traub, W. A., Fazio, G. G., and Low, F. J., 1977. *Bull. Amer. Astron. Soc.* **9** 511.

Taylor, G. E., 1969. *Jnl. Brit. Astron. Assoc.* **80** 13 ff.

Trafton, L., 1974. The source of Neptune's internal heat and the value of Neptune's tidal dissipation factor. *Astrophys. J.* **193** 477.

Wenkert, D. D., Danielson, G. E., and Pollack, J. B., 1984. Imaging of Uranus and Neptune from Voyager 1 and 2, and implications for their internal heat sources. *Bull. Amer. Astron. Soc.* **16** 659.

Wildt, R., 1947. *MNRAS* **107** 84 ff.

Young, C. A., 1891. *General Astronomy.* London and Boston, p. 372.

CHAPTER 7

Andriyanycheva, S. B., Gajkovish, K. P., and Naumov, A. P., 1979. On estimates of ammonia and water vapour content in the atmospheres of Uranus and Neptune. *Izv. vuzov. Radiofiz.* **22** 888 ff. (In Russian.)

Appleby, J. F., 1979. Model atmospheres of Uranus and Neptune. *Bull. Amer. Astron. Soc.* **111** 569.

Appleby, J. F., 1986. Models of Uranus and Neptune, *Icarus* **65** 335.

Apt, J., Clark, R. N., and Singer, R. B., 1980. Photometric spectra of the elusive Neptune haze. *Bull. Amer. Astron. Soc.* **12** 705.

Atai, A. A., and Ibragimov, N. B., 1977. The spectrum of Neptune in the region λλ4400–7000 Å. *Astron. Tsirk.* No. 940, p. 4. (In Russian.)

Atai, A. A., 1980. Uranus and Neptune: spectrophotometry in the region λλ4300–7000 Å. *Astron. Vestn.* **14** 154 ff. (In Russian)

Atai, A. A., 1983. The spectrum of Neptune at λλ0.45–0.70 μm; a comparison with simplest models of the formation of CH_4 absorption bands. *Tsirk. Shemakh. Astrofiz. Obs.* No. 70, pp. 3–6. (In Russian.)

Atai, A. A., and Kyrkhlarov, F. N., 1985. Absorption bands in Neptune's spectrum in the λλ4000–5000 Å region. *Astron. Tsirk.* No 1391, pp. 5–7. (In Russian.)

Atreya, S. K., and Donahue, T. M., 1974. Ionospheric Models for Saturn, Uranus and Neptune. *Icarus* **24** 358.

Atreya, S. K., 1984. Aeronomy. NASA CP-2330, p. 55 ff. (Theoretical models for the upper atmosphere distribution of the neutral and ionospheric species.)

Baines, K. H., Terrile, R. J., Wenkert, D., and Smith, B. A., 1986. Spatial variability of aerosols in the atmosphere of Neptune. *Bull. Amer. Astron. Soc.* **18** 786.

Baines, K. H., Bergstralh, J. T., Terrile, R. J., Wenkert, D., Neff, J., Smith, B. A., and Smith, W. H., 1987. Aerosol and gas distributions in the troposphere of Neptune; constraints from 'visible' broad-band imagery and low- and high-resolution spectrophotometry. *Bull. Amer. Astron. Soc.* **19** 639.

Belton, M. J. S., Wallace, L., and Howard, S., 1981. The periods of Neptune; evidence for atmospheric motions. *Icarus* **46** 263.

Bergstralh, J. T., and Neff, J. S., 1983. Absolute spectrophotometry of Neptune: 3390 to 7800 Å. *Icarus* **55** 40 ff.

Bergstralh, J. T., and Baines, K. H., 1984. Properties of the upper troposphere of Uranus and Neptune derived from observations at 'visible' to near infra-red wavelengths. *Uranus and Neptune* NASA CP-2330, p. 179 ff.

Bergstralh, J. T., Baines, K. H., and Neff, J. S., 1984. Aerosols in the atmospheres of Uranus and Neptune: comparison of their physical properties and distribution. *Bull. Amer. Astron. Soc.* **16** 660.

Bergstralh, J. T., and Baines, K. H., 1985. Aerosols in the atmosphere of Neptune. *Bull. Amer. Astron. Soc.* **17** 744.

Bergstralh, J. T., Baines, K. H., and Neff, J. S., 1986. Aerosols in the stratosphere and high troposphere of Neptune: constraints from near-IR broadband absolute spectrophotometry. *Bull. Amer. Astron. Soc.* **18** 786.

Bergstrain, J. T., Baines, K. H., Terrile, R. J., Wenkert, D., Neff, J., and Smith, B. A., 1987. Aerosols in the stratosphere of Neptune: constraints from near-IR broadband imagery and UV, blue, and near-IR spectrophotometry. *Bull. Amer. Astron. Soc.* **19** 639.

Brown, R. H., Cruikshank, D. P., and Tokunaga, A. T., 1981. The rotation period of Neptune's upper atmosphere. *Icarus* **47** 159 ff.

Caldwell, J., Owen, T., Rivolo, A. R., Moore, V., Hunt, G. E., and Butterworth, P. S., 1981. Observations of Uranus, Neptune and Titan by the International Ultraviolet Explorer. *Astron. J.* **86** 298 ff.

Caldwell, J., Winkelstein, P., Owen, T., Combes, M., Encrenaz, T., Hunt, G., and Moore, V. 1982. Observations of Uranus and Neptune with the International Ultraviolet Explorer. *Advances in ultraviolet astronomy,* p. 306.

Caldwell, J., Wagener, R., and Fricke, K., 1984. Observations of Uranus and Neptune with the IUE satellite. *Bull. Amer. Astron. Soc.* **16** 659.

Caldwell, J., Wagener, R., Owen, T., Combes, M., and Encrenaz, T., 1984. NASA CP-2349, p. 501.

Caldwell, J., Wagener, R., Owen, T., Combes, M., and Encrenaz, T., 1984. Ultraviolet observations of Uranus and Neptune below 3000 Å. *Uranus and Neptune.* NASA CP-2330, p. 157.

Combes, M., Encrenaz, T., Butterworth, P., Owen, T., Caldwell, J., Wagener, R., Hunt, G., and Moore, V., 1983. Search for C_2H_2 in the atmospheres of Uranus and Neptune using the IUE satellite. *Bull. Amer. Astron. Soc.* **15** 858.

Conway, R. G., Davis, R. J., and Padin, S., 1986. Microwave measurements of the outer planets. *MNRAS* **219** 31P–33P.

Courtin, R., Gautier, D., and Lacombe, A., 1979. Indications of supersaturated stratospheric methane in Neptune from its atmospheric thermal profile. *Icarus* **37** 236 ff.

Dementev, M. S., 1982. Molecular absorption in the short-wave spectral region of Neptune. *Astron. Tsirk.* No. 1236, pp. 3–5. (In Russian.)

De Pater, I., and Massie, S. T., 1985. Models of the millimetre–centimetre spectra of the giant planets. *Icarus* **62** 143.

Drobyshevskii, E. M., 1979. On the equatorial currents of Uranus and Neptune. *Astron. Zh.* **56** 1060 ff. (In Russian; English translation in *Soviet Astronomy* **23** No. 5.)

Fink, H., and Larson, P., 1979. The infrared spectra of Uranus, Neptune and Titan from 0.8 to 2.5 microns. *Astrophys. J.* **233** 1021 ff.

French, R. G., Elias, J. H., Mink, D. J., and Elliott, J. L., 1981. The structure of Neptune's upper atmosphere; the stellar occultation of 24 May 1981. *Icarus* **55** 332 ff.

Gautier, D., and Courtin, R., 1979. Atmospheric thermal structures of the giant planets. *Icarus* **39** 28.

Gelfard, J., White, R. E., Perlsweig, D. M. P., and Smith, W. H., 1977. On the 6725 Å band of methane as observed in Uranus and Neptune. *Astrophys. J. Lett.* **215** L43–L45.

Gillett, F. C., and Rieke, G. H., 1977. 5 to 20 micron observations of Uranus and Neptune. *Astrophys. J. Lett.* **218** L141–L144.

Golitsyn, G. S., 1979. Atmospheric dynamics on the outer planets and some of their satellites. *Icarus* **38** 333.

Hammel, H. B., 1986. Methane-band imaging of Neptune and Uranus. *Bull. Amer. Astron. Soc.* **18** 764.

Hardorp, J., 1981. Albedos of Uranus and Neptune and solar color. *Astron. Astrophys.* **96** 123.

Hayden-Smith, W., Schempp, W. V., and Baines, K., 1986. Methane clouds in Neptune's atmosphere. *Bull. Amer. Astron. Soc.* **18** 786.

Herzberg, G., 1951. Spectroscopic evidence of molecular hydrogen in the atmospheres of Uranus and Neptune. *Jnl. RAS of Canada* **45** 100. See also *Astrophys J.* **115** 337 (1952).

Hildebrand, R. H., 1985. Far infra-red and submillimetre brightness temperatures of the giant planets. *Icarus* **64** 64.

Hubbard, W. B., Lellouch, E., Sicardy, B., Bouchet, P., Vilas, F., and Narayan, R., 1986. Effects of Neptunian atmospheric waves on central flash and immersion/emersion occultation profiles. *Bull. Amer. Astron. Soc.* **18** 786.

Hunt, G. E., 1977. Weather on Neptune. *Nature* **269** 10.

Hunten, D. M., 1984. Atmospheres of Uranus and Neptune. *Uranus and Neptune.* NASA CP-2330, 27 ff.

Ibragimov, K. Y., and Sorokina, L. P., 1977. On the methane clouds in the atmospheres of Uranus and Neptune. *Astron. Tsirk.* No. 941, pp. 4–6. (In Russian.)

Ibragimov, K. Y., Kirienko, G. A., and Solodovnik, A. A., 1986. On the possible cloud structure of Uranus and Neptune. *Astron. Tsirk.* No. 1445, pp. 6–8. (In Russian.)

Ingersoll, A. P., 1984. Atmospheric dynamics of Uranus and Neptune: theoretical considerations. *Uranus and Neptune.* NASA CP-2330, 263 ff.

Joyce, R. R., Pilcher, C. B., Cruikshank, D. P., and Morrison, D., 1977. Evidence for weather on Neptune, I. *Astrophys. J.* **214** 657 ff.

Joyce, R. R., Pilcher, C. B., Cruikshank, D. P., and Morrison, D., 1977. Evidence for weather on Neptune, II. *Astrophys. J.* **214**, 663 ff.

Khare, B. N., Sagan, C., Thompson, W. R., Arakawa, E. T., and Votaw, P., 1986. Solid Hydrocarbon Aerosols produced in simulated Uranian and Neptunian stratospheres. Cornell University, *CRSR 869.*

Kovalevsky, J., and Link, K., 1969. *Astron. Astrophys.* **2** 398.

Larson, H. P., 1979. The infra-red spectra of Uranus, Neptune and Titan from 0.8 to 2.5 microns. *Astrophys. J.* **233** 1021.

Lellouch, E., Hubbard, W. B., Sicardy, B., Vilas, F., and Bouchet, P., 1986. Occultation determination of Neptune's oblateness and stratospheric methane mixing ratio. *Nature* **324** 227 ff.

Loewenstein, R. F., Harper, D. A., and Moseley, H., 1977a. Far infrared observations of Neptune. *Bull. Amer. Astron Soc.* **9** 431.

Loewenstein, R. F., Harper, D. A., and Moseley, H., 1977b. The effective temperature of Neptune. *Astrophys. J. Lett.* **218** L145–L146.

Macy, W., and Trafton, L., 1975a. Neptune's atmosphere: concerning the thermal structure. *Bull Amer. Astron. Soc.* **7** 383.

Macy, W., and Trafton, L., 1975b. Mixing ratios of methane, ethane and acetylene in Neptune's stratosphere. *Icarus* **41** 153 ff.

Macy, W., and Smith, W., 1977. The 6819 Å line of Uranus and Neptune. *Bull Amer. Astron. Soc.* **9** 471.

Macy, W., and Sinton, W., 1977a. Ethane and methane emission by Neptune. *Bull Amer. Astron. Soc.* **9** 537.

Macy, W., and Sinton, W., 1977b. Detection of methane and ethane emission on Neptune but not on Uranus. *Astrophys. J. Lett.* **218** L79–L81.

Macy, W., Gelfard, J., and Smith, W. H., 1978. Interpretation of the 6818.9 Å methane line in terms of inhomogenous scattering models for Uranus and Neptune. *Icarus* **34** 20 ff.

Morrison, D., and Cruikshank, D. P., 1972. Temperatures of Uranus and Neptune at 24 microns. *Publ. Astr. Soc. Pacific* **84** 642.

Moseley, H., Conrath, B., and Silverberg, R. F., 1985. Atmospheric temperature profiles of Uranus and Neptune. *Astrophys. J. Lett. Ed.* **292** L83–L86.

Münch, G., and Hippelein, H., 1980. The effects of seeing on the reflected spectrum of Uranus and Neptune. *Astron. Astrophys.* **81** 189 ff.

Neff, J. S., 1984. Absolute spectrophotometry of Titan, Uranus and Neptune at 3500 and 10 500 Å. *Icarus* **60** 221.

Neff, J. S., Bergstralh, J. T., and Baines, K. H., 1986. Raman scattering in the ultraviolet spectra of Uranus and Neptune. *Bull. Amer. Astron. Soc.* **18** 938.

Olsen, E. T., and Gulkis, S., 1978. A preliminary investigation of Neptune's atmosphere via its microwave continuum emission. *Bull. Amer. Astron. Soc.* **10** 577.

Orton, G. S., Tokunaga, A. T., and Caldwell, J., 1981. Evidence for thermal inversions in the lower stratospheres of Uranus and Neptune. *Bull. Amer. Astron. Soc.* **13** 732.

Orton, G. S., Tokunaga, A. T., and Caldwell, J., 1982. Thermal constraints on atmospheric structure of Uranus and Neptune from observations in the 10 μm spectral region. *Bull. Amer. Astron. Soc.* **14** 760.

Orton, G. S., Tokunaga, A. T., and Caldwell, J., 1983. Observational constraints on the atmospheres of Uranus and Neptune from new measurements near 10 μm. *Icarus* **56** 147 ff.

Orton, G. S., Nolt, I. G., Radostitz, J. V., Griffin, M. J., Cunningham, C. T., Ade, P. A. R., Tokunga, A. T., Caldwell, J., and Robson, E. I., 1983. Far-infrared through millimetre observations of Uranus and Neptune. *Bull. Amer. Astron. Soc.* **15** 858.

Orton G. S., and Appleby, J. F., 1984. Temperature structures and infrared-derived properties of the atmospheres of Uranus and Neptune. *Uranus and Neptune.* NASA CP-2330, 89 ff.

Orton G. S., Aitken, D., Smith, C., Roche, P., Caldwell, J., and Snyder, R., 1986. Hydrocarbons in the stratosphere of Uranus and Neptune. *Bull. Amer. Astron. Soc.* **18** 764.

Orton, G. S., Griffin, M. J., Ade, P. A. R., Nolt, I. G., Radostitz, V., Robson, E. I., and Gear, W. K., 1986. Submillimeter and millimeter observations of Uranus and Neptune. *Icarus* **67** 289 ff.

Orton, G. S., Baines, K. H., Bergstralh, J. T., Brown, R. H., Caldwell, J. and Tokunaga, A. T., 1987. Infrared radiometry of Uranus and Neptune at 21 and 32 μm. *Icarus* **69** 230 ff.

Owen, T., Lutz, B. L., Porco, C. C., and Woodman, J. H., 1974. On the identification of the 6420 Å absorption feature in the spectra of Uranus and Neptune. *Astrophys. J.* **189** 379 ff.

Pilcher, C. B., Joyce, R. R., and Cruikshank, D. P., 1977. Evidence for weather on Neptune. *Bull. Amer. Astron. Soc.* **9** 471.

Podolak, M., Giver, L., and Goorvitch, D., 1986. Are the aerosols on Uranus and Neptune composed of methane photopolymers? NASA CP-2441, p. 273.

Pollack, J. B., Rages, K., Bergstralh, J., Baines, K., Wenkert, D., and Danielson, G. E., 1986. Vertical structure of aerosols and clouds in the atmospheres of Uranus and Neptune; implications for their heat budgets. NASA CP-2441, p. 261.

Price, M. J., and Franz, O. G., 1980. Neptune: limb brightening within the 7300 ångström methane band. *Icarus* **41** 430 ff.

Prinn, R. G., 1973. The atmospheres of Uranus and Neptune: a review. *Plan. Space Sci.* **21** 1601 ff.

Rages, K., Veverka, J., Wasserman, L., and Freeman, K. C., 1974. The upper atmosphere of Neptune; an analysis of occultation observations. *Icarus* **23** 59 ff.

Rieke, G. H., and Low, F. J., 1974. Infrared measurements of Uranus and Neptune. *Astrophys. J. Lett.* **193** L147–L148.

Romani, P. N., and Atreya, S. K., 1984. Photochemistry of methane in the atmosphere of Neptune. *Bull. Amer. Astron. Soc.* **16** 660.

Savage, B. D., Cochran, W. D., and Wesselius, P. R., 1980. Ultraviolet albedos of Uranus and Neptune. *Astrophys. J.* **237** 627 ff.

Sicardy, B., Maillard, J. P., and Cruikshank, D. P., 1983. Stratospheric temperature of Neptune from stellar occultation. *Bull. Amer. Astron. Soc.* **15** 858.

Sinton, W. M., Albedo and phase variations of Uranus and Neptune. *Lowell Obs. Bull.* **4** No. 95, p. 93.

Smith, B. A., 1984. Near infrared imaging of Uranus and Neptune. *Uranus and Neptune.* NASA CP-2330, 213 ff.

Smith, W. H., Macy, W., and Pilcher, C. B., 1980. Measurements of the H_2 4–0 quadrupole bands of Uranus and Neptune. *Icarus* **43** 153 ff.

Snyder, R., Caldwell, J., Orton, G., Aitken, D., Smith, C., and Roche, P., 1985. Observations of Uranus and Neptune in the spectral regions 8–13 μm and 17–23 μm. *Bull. Amer. Astron. Soc.* **17** 698.

Stier, M. T., Traub, W. A., Fazio, G. G., Wright, E. L., and Low, F. J., 1978. Far-infrared observations of Uranus, Neptune and Ceres. *Astrophys. J.* **226** 347.

Taylor, G. E., 1970. The occultation of BD −17°4388 by Neptune on 1986 April 7. *MNRAS* **147** 27 ff.

Tejfel, V. G., 1985. Clouds on Uranus and Neptune. *Zemlya Vselennaya* No. 5, p. 22. (In Russian.)

Tokunaga, A. T., Orton, G. S., and Caldwell, J., 1983. New observational constraints on the temperature inversions of Uranus and Neptune. *Icarus* **53** 141 ff.

Trafton, J., 1972. On the methane opacity for Uranus and Neptune. *Astron. J. Letters* **172** L117.

Trafton, L., 1974. Neptune: observations of the quadrupole lines in the (4–0) band. In: *Exploration of the Solar System* (Woszczyk and Iwaniszewska eds.), p. 497.

Van Hemelrijck, E., 1982. The oblateness effect of the solar radiation incident at the top of the atmosphere of the outer giants. *Icarus* **51** 39.

Vdovichenko, V. D., Gajsin, S. M., Ilin, V. V., and Mosina, S. A., 1986. Geometrical albedo of Uranus and Neptune in the spectral region λ0.32–0.72 μm. *Astron. Tsirk.* No. 1445, pp. 4–6. (In Russian.)

Veverka, J., Wasserman, L., and Sagan, C., 1974. On the Upper Atmosphere of Neptune. *Astrophys. J.* **189** 569.

Wagener, R., Caldwell, J., and Fricke, K., 1986. The geometrical albedos of Uranus and Neptune between 2100 and 3350 Å. *Icarus* **67** 281 ff.

Wallace, L., 1984. The seasonal variation of the thermal structure of the atmosphere of Neptune. *Icarus* **59** 367 ff.

Warnsteker, W., 1973. The wavelength dependence of the albedos of Uranus and Neptune from 0.3 to 1.1 microns. *Astrophys. J.* **184** 1007.

Weidenschilling, S. J., and Lewis, J. S., 1973. Atmospheric and cloud structures of the Jovian planets. *Icarus* **20** 465 ff.

Whitcomb, S., 1979. Sub-millimetre brightness temperatures of Venus, Jupiter, Uranus and Neptune. *Icarus* **38** 75.

Wildt, R., 1937. Decomposition of methane in the atmospheres of Uranus and Neptune. *Astrophys. J.* **86** 321. (See also Adel and Slipher, *Phys. Rev.* (2) **47** 787.)

Zharkov, V., 1972. Adiabatic temperatures in Uranus and Neptune. *Izv. Akad. Nauk. SSR. Fiz Zemli* **7** 120–127.

CHAPTER 8

Berge, C. L., 1968. Recent observations of Saturn, Uranus and Neptune at 3.12 cm. *Astrophys. J. Lett.* **2** 127.

Curtis, S. A., and Ness, J. F., 1986. Magnetospheric balance in planetary dynamos: predictions for Neptune's magnetosphere. *J. Geophys. Res.* **91** No. A10, pp. 11003–11008.

De Pater, I., and Richmond, M., 1988. *Sky and Telescope* **76,** 343.

Hill, T. W., 1984. Magnetospheric structures: Uranus and Neptune. *Uranus and Neptune.* NASA CP-2330, 497 ff.

Hunt, G. E., and Moore, P., 1988. *Atlas of Uranus.* Cambridge.

Kavanagh, L. D., 1975. Synchroton radio emission from Uranus and Neptune. *Icarus* **25** 166 ff.

Kellerman, K. L., and Pauliny-Toth, I. I. K., 1966. Observation of the radio emission of Uranus, Neptune and Pluto at wavelengths of 11.3 and 3.7 cm. *Astrophys. J.* **145** 954.

Kellerman, K. L., and Pauliny-Toth, I. I. K., 1970. Millimetre wavelength measurements of Uranus and Neptune. *Astrophys. J. Lett.* **6** 185.

Kennel, C. F., and Maggs, J. E., 1976. Possibility of detecting magnetospheric radio bursts from Uranus and Neptune. *Nature* **261** 299.

Mayer, C. H., and McCullough, T. P., 1971. Microwave radiation of Uranus and Neptune. *Icarus* **14** 187 ff.

Siscoe, G. L., 1979. Towards a comparative theory of magnetospheres. In: *Solar*

System Plasma Physics — a Twentieth Century Review (Kennel, C. F., Lanzer-otti, J., and Parker, E. N. eds).

Smith, A. G., Leacock, R. J., Carr, T. D., Leno, G. R., and Olsson, C. N., 1973. New Thermal Radio Observation of the Giant Planets. *Phys. Earth. Planet. Inter.* **6** 10.

Smoluchowski, R., and Torbett, M., 1981. Can magnetic fields be generated in the icy mantles of Uranus and Neptune? *Icarus* **48** 146.

Stevenson, D., 1974. Origin and maintenance of the magnetic field: a precessionally-driven dynamo? *Icarus* **22** 403.

Torbett, M., and Smoluchowski, R., 1980. Hydrodynamic dynamo in the cores of Uranus and Neptune. *Nature* **286** 237.

Webster, W. J., Webster, A. C., and Webster, G. T., 1972. Interferometer observations of Uranus, Neptune and Pluto at wavelengths of 11.1 and 3.7 cm. *Astrophys. J.* **174** 679.

CHAPTER 9

Allen, D. A., 1983. Infrared views of the giant planets. *Sky and Tel.* **65** 112.

Allen, D. A., 1984. An infra-red Astronomer looks at cloud-covered planets. In: *Yearbook of Astronomy 1985* (Moore, P. ed.), p. 138.

Borderies, N., Goldreich, P., and Tremaine, S., 1986. An explanation for Neptune's arc rings. *Bull. Amer. Astron. Soc.* **18** 778.

Brahic, A., Sicardy, B., Roques, F., McLaren, R., and Hubbard, W. B., 1986. Neptune's arcs: where and how many? *Bull. Amer. Astron. Soc.* **18** 778.

Covault, C. E., Glass, I. S., French, R. G., and Elliott, J. L., 1986. The 7 and 25 June 1985 Neptune occultations: constraints on the putative Neptune 'arc'. *Icarus* **67** 126 ff.

Dobrovolskis, A. R., 1980. Where are the rings of Neptune? *Icarus* **43** 222.

Dunham, D. W., 1978. Occultations by Neptune and the rings of Uranus. *Occultation Newsletter* **1** 131.

Dzērvitis, U., 1983. Has Neptune also a ring? Zvaigžnota Debess, gada pavasaris, pp. 14–16. (In Latvian.)

Elliot, J. L., Mink, D. J., Elias, J. H., Baron, R. L., Dunham, E., Pingree, J. E., French, R. G., Liller, W., Nicholson, P. D., Jones, T. J. and Franz, O. G., 1981. No evidence of rings around Neptune. *Nature* **294** 526 ff.

Elliot, J. L., 1982. Rings around Neptune? *Bull. Amer. Astron. Soc.* **14** 750.

Elliot, J. L., 1983. *IAU Circular,* No. 3831.

Elliot, J. L., Dunham, E., Mink, D. J., Meech, K. J., Goguen, J., Hammel, H. B., and Erickson, E. F., 1983. Occultation limits on rings around Neptune from Mauna Kea and the Kuiper Airborne Observatory. *Bull. Amer. Astron. Soc.* **15** 817.

Elliot, J. L., 1984. Rings around Neptune? Planetary rings, p. 197–199. *IAU Colloq.* No. 75.

Elliot, J. L., 1984. The structure of the Uranian rings and the search for rings around Neptune. *Uranus and Neptune.* NASA CP-2330, 575–588.

Elliot, J. L., and Kerr, R., 1984. *Rings,* pp. 179–186. Cambridge (Massachusetts) and London.

Elliot, J. L., Baron, R. L., Dunham, E. W., French, R. G., Meech, K. J., Mink, D.

J., Allen, D. A., Ashley, M. C. B., Freeman, K. C., Erickson, E. F., Goguen, J., and Hammel, H. B., 1985. The 1983 June 15 occultation by Neptune. I. Limits on a possible ring system. *Astron. J.* **90** 2615 ff.

Goldreich, P., Tremaine, S., and Borderies, N., 1986. Towards a theory for Neptune's arc rings. *Astron. J.* **92** 490 ff.

Guinan, E. F., Harris, C. C., and Maloney, F., 1968. Occultation of BD −17°4388 on 7 April, 1968. *Bull. Amer. Astron. Soc.* **14** 658.

Guliev, A. S., 1980. Conditions for the visibility of a possible ring of Neptune. *Komet. Tsirk.* (Kiev) No. 259. (In Russian.)

Häfner, R., Manfroid, J., and Bouchet, P., 1985. Discovery of Neptune's ring at La Silla. *Messenger* No. 42, 10–12.

Häfner, R., and Manfroid, J., 1985. Entdeckung eines Rings um Neptun. *Sterne Weltraum* **24** Jahrg. Nr. 7, 382–384. (In German.)

Hubbard, W. B., Frecker, J. E., Gehrels, J. A., Gehrels, T., Hunten, D. M., Lebofsky, L. A., Smith, B. A., Tholen, D. J., Vilas, F., Zellner, B., Avey, H. P., Mottram, K., Murphy, T., Varnes, B., Carter, B., Nielsen, A., Page, A. A., Fu, H. H., Wu, H. H., Kennedy, H. D., Waterworth, M. D., and Reitsema, H. J., 1985. Results from the observations of the 15 June 1983 occultation by the Neptune system. *Astron. J.* **90** 655 ff.

Hubbard, W. B., Brahic, A., Sicardy, B., Elicer, L. R., Roques, F., and Vilas, F., 1986. Occultation detection of a Neptunian ring-like arc. *Nature* **319** 636 ff.

Hubbard, W. B., 1986. 1981 N1: a Neptune arc? *Science* **231** 1276–1278.

Kerr, R. A., 1981. Neptune's rings fading. *Science* **213** 1240.

Kerr, R. A., 1983. Neptune ring fades again. *Science* **222** 311.

Kerr, R. A., 1984. What's going on at Neptune? *Science* **227** 734.

Lissauer, J. J., 1985. Shepherding model for Neptune's ring. *Nature* **318** 544.

Lissauer, J. J., 1985. A shepherding model for Neptune's arc ring. *Bull. Amer. Astron. Soc.* **17** 719.

Maddox, J., 1985. Whose rings around Neptune? *Nature* **318** 505.

Manfroid, J., Häfner, R. and Bouchet, P., 1985. Qu'y a-t-il autour de Neptune? *Ciel* **48** 212.

Manfroid, J., Häfner, R., and Bouchet, P., 1986. New evidence for a ring around Neptune. *Astron. Astrophys.* **157** No. 1, pp. L3–L5.

Mink, D. J., Klemola, A. R., and Elliot, J. *Astron. J.* **86** 135.

Murray, C. D., 1986. Arcs around Neptune? *Nature* **324** 209.

Pandey, A. K., Mahra, H. S., and Mohan, V., 1984. Occultation of MKE 31 on 12 September 1983. *Earth, Moon and Planets* **31** 217.

Pandey, A. K., and Mahra, H. S., 1987. Possible ring system of Neptune. *Earth, Moon and Planets* **37** 147 ff.

Reitsema, H., Hubbard, W. B., Lebofsky, L., and Tholen, D., 1981. Neptune's Third Satellite? *Sky and Tel.* **62** 317.

Schmidt, M. J., 1985. Auch Planet Neptun bestizt einen Ring. *Orion* **43** Jahrg. Nr. 208, 85. (In German.)

Sicardy, B., Roques, F., Brahic, A., Maillard, J. P., Cruikshank, D. P., and Becklin, E. E., 1983. Limit of detection of faint rings around Uranus and Neptune from stellar occultations. *Bull. Amer. Astron. Soc.* **15** 816.

Sicardy, B., Roques, F., Brahic, A., Bouchet, P., Maillard, J. P., and Perrier, C., 1986. More dark matter around Uranus and Neptune? *Nature* **320** 729.

Thomsen, D. E., 1982. Neptune's rings: an occultation story. *Science* **222** 311.

Vilas, F., Hubbard, W. B., Frecker, J. E., Hunten, D. M., Gehrels, T., Lebofsky, L. A., Smith, B. A., Tholen, D. J., Zellner, B. H., Wisniewsky, W., Gehreis, J. A., Capron, B. 1983. *Bull. Amer. Astron. Soc.* **15** 816.

Vilas, F., and Hubbard, W. B., 1985. The 1985 August 20 Neptune occultation observed from CTIO. *Bull. Amer. Astron. Soc.* **17** 923.

Wilford, J. N., 1982. Data show two rings circling Neptune. *New York Times,* June 10 1982, B. 10.

CHAPTER 10

Alden, H. L., 1940. Mass of the satellite of Neptune. *Astron. J.* **49** 71.

Alden, H. L., 1943. Observations of the satellite of Neptune. *Astron. J.* **50** 110.

Apt., J., Carleton, N. P., and Mackay, C. D., 1983. Methane on Triton and Pluto: new CCD spectra. *Astrophys. J.* **270** 342 ff.

Barnard, E. E., 1910. Observations of the satellite of Neptune. *Nature* **84** 472.

Benner, D. C., Fink, U., and Cromwell, R. H., 1978. Image tube spectra of Pluto and Triton from 6800 to 9000 Å. *Icarus* **36** 82 ff.

Bond, W. C., 1850. Observations of Lassell's satellite of Neptune. *MNRAS* **8** 9.

Bonneau, D., and Foy, F., 1986. First direct measurements of the diameters of the large satellites of Uranus and Neptune. *Astron. Astrophys.* **161** No. 1, L12–L13.

Bronshten, V. A., 1979. Riddling Triton. *Zemlya i Vselennaya* No. 2, 47. (In Russian.)

Brown, R. H., and Cruikshank, D. P., 1985. The moons of Uranus, Neptune and Pluto. *Sci. American* **253** No. 1, 28 ff.

Challis, J., and Thompson, R. A., 1848. Elements of the satellite of Neptune. *MNRAS* **6** 128, 201.

Cochran, W. D., and Cochran, A. L., 1978. Digicon spectroscopy of Triton and Pluto. *Bull. Amer. Astron. Soc.* **10** 585.

Cruikshank, D. P., Pilcher, C. B., and Morrison, D., 1977. Identification of a new class of satellites in the outer Solar System. *Astrophys. J.* **217** 1006 ff.

Cruikshank, D. P., and Silvaggio, P., 1978. Methane atmospheres of Triton and Pluto. *Bull. Amer. Astron. Soc.* **10** 578.

Cruikshank, D. P., and Silvaggio, P., 1979. Triton: a satellite with atmosphere. *Astrophys. J.* **233** 1016 ff.

Cruickshank, D. P., Stockton, A., Dyck, H. M., Becklin, E., and Macy, W., 1979. The diameter and reflectance of Triton. *Icarus* **40** 104 ff.

Cruikshank, D. P., 1982. Spectroscopy of Triton and Pluto: current status and prospects. *Vibrational–rotational spectroscopy for planetary atmospheres.* p. 699 ff.

Cruikshank, D. P., Brown, R. H., and Clark, R. N., 1983. Nitrogen on Triton. *Bull. Amer. Astron. Soc.* **15** 857.

Cruikshank, D. P., 1984. Physical properties of the satellites of Neptune. *Uranus and Neptune.* NASA CP-2330, 425 ff.

Cruikshank, D. P., Brown, R. H., and Clark, R. N., 1984. Nitrogen on Triton. *Icarus* **58** 293 ff.

Cruikshank, D. P., and Apt, J., 1984. Methane on Triton: physical state and distribution. *Icarus* **58** 306 ff.

Cruikshank, D. P., Brown, R. H., and Clark, R.N., 1985. Methane ice on Triton and Pluto. *Ices in the Solar System.* p. 817ff.

Cruikshank, D. P., Brown, R. H., Smith, R. G., and Tokunaga, A. T., 1986. Infrared spectral variations on Triton. *Bull. Amer. Astron. Soc.* **18** 762.

Combes, M., Encrenaz, L., Lecacheux, J., and Perrier, C., 1981. Upper limit of the gaseous CH_4 abundance on Triton. *Icarus* **47** 139.

Davies, M. E., Abalakin, V. K., Burša, M., Hunt, G. E., Lieske, J. H., Morando, B., Rapp, R. H., Seidelmann, P. K., Sinclair, A. T., and Tjuflin, Y. S., 1988. Report of the IAU/IAG/COSPAR Working Group on Cartographic Coordinates and Rotational Elements of the Planets and Satellites. Presented at the IAU Assembly at Baltimore, August 1988.

Delitsky, M. L., 1983. Chemistry of Triton's ocean. *Bull. Amer. Astron. Soc.* **15** 857.

Delitsky, M. L., and Thompson, W. R., 1987. Chemical processes in Triton's atmosphere and surface. *Icarus* **70** 354 ff.

Dermott, S. F., 1984. Origin and evolution of the Uranian and Neptunian satellites: some dynamical considerations. *Uranus and Neptune.* NASA CP-2330, p. 377.

Eichelberger, W., 1926. (Orbital considerations.) *Nature* **117** 428.

Fouche, M., 1905. (Name of Triton) *Nature* **73** 182.

Franz, O. G., 1981. UBV photometry of Triton. *Icarus* **45** 602 ff.

Gill, J. R., and Gault, B. L., 1968. A new determination of the orbit of Triton, pole of Neptune's equator, and mass of Neptune. *Astron. J.* **73** 895.

Golitsyn, G. S., and Steklov, A. F., 1981. On the atmospheres of Triton and Pluto. *Physics of Planetary Atmospheres.* p. 139 ff. (In Russian.)

Greenberg, R., 1984. Satellite masses in the Uranus and Neptune systems. *Uranus and Neptune.* NASA CP-2330, 463 ff.

Greene, T. F., Johnson, P. E., and Shorthill, R. W., 1976. The spectral reflectivities of Ariel, Umbriel, Titania, Oberon and Triton. *Bull. Amer. Astron. Soc.* **8** 464.

Harris, A. W., 1981. A redetermination of the orbit of Triton. *Bull. Amer. Astron. Soc.* **13** 573.

Harris, A. W., 1982. The effect of Triton's mass on the determination of the J_2 of Neptune. *Bull. Amer. Astron. Soc.* **14** 580.

Harris, A. W., 1983. A redetermination of the orbit of Triton. *Bull. Amer. Astron. Soc.* **13** 573.

Harris, A. W., 1983. A redetermination of the orbit of Triton. *Bull. Amer. Astron. Soc.* **15** 870.

Harris, A. W., 1984a. Physical properties of Neptune and Triton inferred from the orbit of Triton. *Uranus and Neptune.* NASA CP-2330, 357 ff.

Harris, A. W., 1984b. Triton not doomed. *Sky and Tel.* **67** 108.

Hind, J. R., 1955. On the satellite of Neptune. *MNRAS* **15** 46.

Jeličič, M., 1981. The discovery of the third satellite of Neptune. *Vasiona* Année **29**, 62. (In Serbo-Croatian.)

Johnson, J. R., Fink, U., Smith, B. A., and Reitsema, H. J., 1981. Spectrophotometry and upper limit for gaseous CH_4 on Triton. *Icarus* **46** 288 ff.

Johnson, P. E., Greene, T. F., and Shorthill, R. W., 1978. Narrow-band spectrophotometry of Ariel, Umbriel, Titania, Oberon and Triton. *Icarus* **36** 75 ff.

Johnson, P. E., 1979. Narrow-band spectrophotometry of Ariel, Umbriel, Titania, Oberon and Triton. *Icarus* **39** 139. (Erratum.)

Kostinsky, M. S., 1900. Observations of the satellite of Neptune. *Nature* **62** 161.

Kuiper, G., 1949. (Discovery of Nereid.) *Publ. Astron. Soc. Pacific* **61** 175.

Kuiper, G., 1954. (Diameter of Triton.) *Transactions IAU* **9** 250.

Lassell, W., 1846. (Discovery of the satellite.) *MNRAS* **7** 167.

Lassell, W., 1847. Observations of Neptune and its satellite. *MNRAS* **7** 30.

Lassell, W., 1850. (Letter to Airy concerning a possible second satellite.) 14 August 1850.

Lassell, W., 1852. (Possible second satellite.) *MNRAS* **12** 155.

Lebofsky, L. A., Rieke, G. H., and Lebofsky, M. J., 1982. The radii and albedos of Triton and Pluto. *Bull. Amer. Astron. Soc.* **14** 766.

Lewis, J. S., 1971. Satellites of the outer planets; their physical and chemical nature. *Icarus* **15** 174.

Lewis, J., Bryant, J., and Bowyer, J., 1900. Micrometric measures of the diameter of Neptune, and the distance and position angle of the satellite, made with the 28-inch refractor at the Royal Greenwich Observatory. *MNRAS* **61** 9.

Liu, L., Zhou, Q., and Li , L., 1983. Approach to the theory of motion of Nereid. *Acta Astron. Sin.* **24** No. 1 80 ff. (In Chinese.)

Lohse, J., 1887. The magnitude of Triton. *MNRAS* **47** 497.

Lunine, J. I., and Stevenson, D. J., 1985. Physical state of volatiles on the surface of Triton. *Nature* **317** 238.

Lutz, B. L., Owen, T., and Case, R. D., 1976. Laboratory band strengths of methane and their application to the atmospheres of Jupiter, Saturn, Uranus, Neptune and Titan. *Astrophys. J.* **203** 541.

McCord, T. B., 1966. Dynamical evolution of the Neptunian system. *Astron. J.* **71** 585.

Marth, A., 1878–1881. Ephemeris of the Satellite of Neptune. *MNRAS* **38** 475; **41** 415; **45** 63; **46** 504; **47** 574; **48** 410: **50** 558; **51** 563.

Mignard, F., 1975. Satellite a forte excentricité. Application à Nereide. *Astron. Astrophys.* **43** 359. (In French.)

Mignard, F., 1981. The mean elements of Nereid. *Astron. J.* **86** 1728.

Morrison, D., Cruikshank, D. P., and Brown, R. H., 1982. Diameters of Triton and Pluto. *Nature* **300** 425.

Newton, H. W., 1922. (Satellite observations and oblateness.) *Nature* **109** 528.

Nicholson, S. B., 1931. The mass of Triton. *Publ. Astron. Soc. Pacific* **43** 261.

Peale, S. J., 1977. (Variations of Triton.) In: *Planetary Satellites* (Burns, J. ed.), Arizona, p. 87.

Perrine, C. D., 1903. The satellite of Neptune. *Nature* **68** 353.

Pickering, W., 1879. The diameter of the satellite of Neptune. *Cambridge Ann.* **11** Ch. 10.

R.G.O., 1903. Observations of the satellite of Neptune, from photographs taken at the Royal Greenwich Observatory between 1902 November 12 and 1903 April 27. *MNRAS* **63** 503. See also *MNRAS* **64** 835 (1904); **66** 10 (1905); **67** 90 (1906);

68 33, 586 (1907); and discussion between Dyson, F. W., and Edney, D. J. R., **65** 570—583 (1905).

Reitsema, H. J., Hubbard, W. B., Lebofsky, L. A., and Tholen, D. J., 1982. Occultation by a possible third satellite of Neptune. *Science* **215** 289 ff.

Riedinger, J., 1984. Triton atmet. *Stern Weltraum* **23** Jahrg. Nr. 1, 4.

Rieke, G. H., Lebofsky , L. A., Lebofsky, M. J., and Montgomery, E. F., 1981. Unidentified features in the spectrum of Triton. *Prepr. Steward Obs.* No. 333, 5 pp.; and *Nature* **294** 59.

Rieke, G. H., Lebofsky, L. A., and Lebofsky, M. J., 1985. A search for nitrogen on Triton. *Icarus* **64** 153 ff.; and *Prepr. Steward Obs.* No. 587, 7 pp.

Roberts, I., 1891. Photographs of Neptune and its satellite. *MNRAS* **51** 439.

Roques, F., Sicardy, B., Bouchet, P., Brahic, A., Häfner, R., Lecacheux, J., and Manfroid, J., 1984. Possible detection of a 10-km-sized object around Neptune. *Bull. Amer. Astron. Soc.* **16** 1027.

Rose, L. E., 1974. Orbit of Nereid and the mass of Neptune. *Astron. J.* **79** 489.

Russell, H. N., 1932. (Mass, density and name of Triton.) *Nature* **129** 405.

Schaeberle. J., 1895. Possible new satellite of Neptune. *Astron. J.* **15** No. 340, 26.

Schaefer, M. W., and Schaefer, B. E., 1988. Large-amplitude photometric variations of Nereid. *Nature* **333** 436.

Sky and Telescope, 1988. Triton's mysterious surface: solid or slushy? *Sky and Tel.* **75** 580.

Spinrad, H., 1969. Lack of a noticeable methane atmosphere on Triton. *Publ. Astron. Soc. Pacific* **81** 895.

Stevenson, D. J., 1984. Composition, structure and evolution of the Uranian and Neptunian satellites. *Uranus and Neptune.* NASA CP-2330, 405 ff.

Strazzulla, G., Calcagno, L., and Foti, G., 1984. Build-up of carbonaceous material by fast protons on Pluto and Triton. *Astron. Astrophys.* **140** 441 ff.

Struve, O., 1894. (Acceleration of Triton.) *Nature* **49** 324.

Tokunaga, A. T., Smith, R. G., Brown, R. H., Cruikshank, D. P. and Clark, R. N., 1985. The infrared spectrum of Triton at 2.0–2.25 μm. *Bull. Amer. Astron. Soc.* **17** 698.

Trafton, L. M., 1984a. Seasonal variations in Triton's atmospheric mass and composition. *Uranus and Neptune.* NASA CP-2330, 481 ff.

Trafton, L., M., 1984b. Large seasonal variations in Triton's atmosphere. *Icarus* **58** 312 ff.

Trafton, L. M., 1985. Triton's seasonal atmospheric changes: on the influence of surface heat conduction. *Bull. Amer. Astron. Soc.* **17** 742.

van Flandern, T., and Harrington, R. S., 1978. A dynamical study of escaped satellites of Neptune. *Bull. Amer. Astron. Soc.* **9** 621.

Veillet, C., 1982. Orbital elements of Nereid from new observations. *Astron. Astrophys.* **112** 277 ff.

Veverka, J., 1988. Taking a dim view of Nereid. *Nature* **333**, 394.

CHAPTER 11

Boynton, J., 1922. Neptune: projected evolution of Triton's orbit. *Publ. Astron. Soc. Pacific* **94** 749.

Čelebonović, V., 1986. On the origin of Triton. *Earth, Moon and Planets* **34** 59 ff.

Cohen, C. J., and Hubbard, E. C., 1964. Libration of Pluto–Neptune. *Science* **145** 1302.

Cohen, C. J., and Hubbard, E. C., 1965. Libration of the close approaches of Pluto to Neptune. *Astron. J.* **70** 10.

Cruikshank, D. P., 1984. Physical Properties of the Satellites of Neptune. *Uranus and Neptune.* NASA CP-2330, 425 ff.

Dauvillier, A., 1951. The nature of Pluto and Triton. *Comptes rendus* **233** 901.

Dermott, S. F., 1984. Origin and evolution of the Uranian and Neptunian satellites: some dynamical considerations. *Uranus and Neptune* NASA CP-2331, 377 ff.

Dormand, J. R., and Woolfson, M. M., 1980. The origin of Pluto. *MNRAS* **193** 171 ff.

Duncombe, R. L., Klepczynski, W., and Seidelmann, P. K., 1988. Orbit of Neptune and the mass of Pluto. *Astron. J.* **73** 830.

Duncombe, R. L., and Seidelmann, P. K., 1980. A history of the determination of Pluto's mass. *Icarus* **44** 12 ff.

Farinella, P., Milani, A., Nobili, A. M., and Valsecchi, G. B., 1979. Tidal evolution and the Pluto–Charon system. *Earth, Moon and Planets* **20** 415 ff.

Harrington, R. S., and van Flandern, T. C., 1979. The satellites of Neptune and the origin of Pluto. *Icarus* **39** 131.

Harrington, R. S., and Harrington, B. J., 1979. The discovery of Pluto's moon. *Mercury* **8** 1.

Hughes, D. W., 1988. Interior and origin of Pluto. *Nature* **335**, 205.

Jackson, J., 1930. Neptune's orbit. *MNRAS* **90** 728.

Kinoshita, H., and Nakai, H., 1984. Motions of the perihelions of Neptune and Pluto. *Celest. Mech.* **34** Nos. 1–4, 203–217.

Klemola, A. R., and Harlan, E. A., 1982. Astrometric observations of outer planets and minor planets: 1980–1982. *Astron. J.* **87** 1242; and *Lick Obs. Bull.* No. 925.

Kuiper, G. P., 1956. The Formation of the Planets. *Jnl. RAS of Canada* **50** 171 ff.

Kuiper, G. P., 1957. Further studies on the origin of Pluto. *Astrophys. J.* **125** 287.

Lin, D. N. C., 1981. On the origin of the Pluto–Charon system. *MNRAS* **197** 1081 ff.

Luyten, W. J., 1956. Pluto not a Planet? *Science* **123** 896.

Lyttleton, R. A., 1936. On the possible results of an encounter of Pluto with the Neptunian system. *MNRAS* **97** 108 ff.

McCord, T. B., 1966. Dynamical evolution of the Neptunian system. *Astron. J.* **71** 585.

McKinnon, W. B., 1982. On the origin of Triton. *Bull. Amer. Astron. Soc.* **14** 765.

McKinnon, W. B., 1984. On the origin of Triton and Pluto. *Nature* **311** 355 ff.

McKinnon, W. B., and Mueller, S., 1988. Pluto's structure and composition suggest origin in the solar, not a planetary, nebula. *Nature* **335**, 240.

Meeus, J., 1972. Résonances dans le système Neptune–Pluton. *L'Astronomie* **86** 33 ff.

Mignard, F., 1981. On a possible origin of Charon. *Astron. Astrophys.* **96** L1–L2.

Nacozy, P. E., and Diehl, R. E., 1978. A semianalytical theory for the long-term motion of Pluto. *Astron. J.* **83** 522 ff.

Rapaport, M., Requième, Y., Mazurier, J. M., and Francou, G., 1987. Meridian observations of Uranus and Neptune at Bordeaux Observatory. Comparison with ephemerides. *Astron. Astrophys.* **179** 317 ff.

Seidelmann, P. K., Duncombe, R. L., and Klepczynski, W. J., 1969. The mass of Neptune and the orbit of Uranus. *Astron. J.* **74** 776.

Seidelmann, P. K., Kaplan, G. H., Pulkkinner, K. F., Santoro, E. J., and van Flandern, T., 1980. *Icarus* **44** 19.

Sessin, W., and Tsuchida, M., 1983. Commensurability in the Uranus–Neptune system. *The motion of planets and natural and artificial satellites.* p. 263 ff.

Stevenson, D. J., 1984. Composition, structure and evolution of the Uranian and Neptunian satellites. *Uranus and Neptune.* NASA CP-2330, 405 ff.

Tombaugh, C. W., and Moore, P., 1981. *Out of the Darkness: the Planet Pluto.* Guildford and Harrisburg.

van Flandern, T., 1986. Personal communication.

Williams, J. G., and Benson, G. S., 1971. Resonances in the Neptune–Pluto system. *Astron. J.* **76** 167.

Wylie, I. R., 1942. A comparison of Newcomb's tables of Neptune with observation, 1795–1938. *Publ. US Naval Obs.* **15** part 1.

CHAPTER 12

Anderson, J. D., and Standish, E. M., 1986. Dynamical evidence for Planet X. *The Galaxy and the Solar System.* p. 286 ff.

Anderson, J. D., 1987. In: *Aviation Week and Space Technology,* 6 July, p. 32.

Anderson, J. D., 1987. *NASA News,* Release 87–32 and 87–35, 26 June.

Beesley, D. E., 1973. The distance and period of a Transplutonian planet as based on the Titius–Bode Law. *Irish Astron. J.* **11** 138.

Brady, J. L., 1972. The effect of a Trans-Plutonian planet on Halley's Comet. *Publ. Astron. Soc. Pacific* **84** 314.

Chebotarev, G. A., 1972. *Byull. Inst. Teoret. Astron.* (Leningrad) **13** 145. (In Russian.)

Chebotarev, G. A., 1975. Associated Press dispatch, Moscow, published in the Arizona (Phoenix) *Republic,* 8 June 1975.

Dallet, G., 1901. Contribution à la recherche des planètes située au delà l'orbite de Neptune. *Bull. Soc. Astron. de France* **15** 266 ff.

Drobyshevskii, E. M., 1976. On Trans-Neptunian Planets. *Astron. Tsirk.* No. 919, 3–5. (In Russian.)

Esclangon, E., 1930. The Trans-Neptunian Planet. *Comptes rendus* **190** 834, 895, 957.

Fernandez, J. A., 1976. Evolution of comet orbits under the perturbing influence of giant planets and nearby stars. *Icarus* **42** 406.

Forbes, G., 1880. On comets and ultra-Neptunian planets. *Observatory* **3** 439.

Forbes, G., 1881. *Proc. Roy. Soc. Edinburgh* **12**.

Foss, A. P. O., and Shawe-Taylor, J. S., 1972. Search for a trans-Neptunian planet. *Nature* **230** 266.

Garnowsky, A., 1902. Sur l'éxistence de quatres planètes transneptunienne. *Bull. Astron. Soc. de France* **16** 484.

Goldreich, P., and Ward, W. R., 1972. The case against Planet X. *Publ. Astron. Soc. Pacific* **84** 737.

Grigull, T., 1902. *Nature* **66** 614.

Guliev, A. S., 1987. On a possibility of existence of a hypothetical planet in the region between Neptune and Pluto. *Kinematika Fiz. Nebsen. Tel.* Tom **3** No. 2, 28 ff. (In Russian. English translation in *Kinematics Phys. Cel. Bodies.*)

Harrington, R. S., and Harrington, B. J., 1979. *Mercury* **8** 1.

Hoyt, P., 1980. *Planets X and Pluto.* Arizona.

Hughes, D. W., 1981. Planet X: is it necessary? *Nature* **291** 613.

Kiang, T., and Wayman, P. A., 1973. The orbit of Halley's Comet. *Nature* **241** 520.

Klemola, A. R., and Harlan, D. A., 1972. Search for Brady's hypothetical trans-Plutonian planet. *Publ. Astron. Soc. Pacific* **84** 736.

Kritzinger, H. H., 1954. Transpluto, hypothetische Elemente. *Nachrichtenblatt der Astronomische Zentralstelle* **8** 4.

Kritzinger, H. H., 1957. *Nachrichtenblatt der Astronomische Zentralstelle* **11** 4.

Lau, H. E., 1900. Planètes inconnues. *Bul. Soc. Astron. de France* **14** 340.

Lau, H. E., 1903. No planet beyond Neptune? *Observatory* **25** 365.

Matese, J. J., and Whitmire, D. P., 1986. Planet X and the origins of shower and steady state flux of short-period comets. *Icarus* **65** 37 ff.

Matese, J. J., and Whitmire, D. P., 1986b. Planet X as the source of periodic and steady-state flux of short-period comets. In: *The Galaxy and the Solar System,* 297 ff.

Meeus, J., 1973. La planète X n'existe pas. *L'Astronomie* **87** 311. (In French.)

Mendez, R. H., 1973. The planet X: putting things in perspective. *Rev. Astron.* **45** Nos. 185–186, 14–16. (In Spanish.)

Moore, P., 1981. Some Thoughts on Planet X. *J. Brit. Astron. Assoc.* **91** 483 ff.

Naef, R., 1955. Hypothetische Elemente eines Transpluto. *Orion* **4** 484. (In German.)

NASA News, 1982. 17 June 1982.

Napter, W., and Dodd, R. J., 1973. The Missing Planet. *Nature* **242** 250.

Naumenko, B. N., Nejman, V. B., Chernov, V. M. and Shapiro, B. V., 1982. On Trans-Plutonian planets in the Solar System. *Astron. Tsirk.* No. 1216 6–8. (In Russian.)

Olssen-Steel, D., 1988. Paper presented at the IAU General Assembly.

Rawlins, D., and Hammerton, M., 1972. Is there a tenth planet in the Solar System? *Nature* **240** 457.

Richardson, R. S., 1942. An attempt to determine the mass of Pluto from its disturbing effect on Halley's Comet. *Publ. Astron. Soc. Pacific* **54** 19 ff.

Roberts, I., 1892. Photographic search for a planet beyond the orbit of Neptune. *MNRAS* **52** 50.

Schütte, K., 1950. In: Kritzinger (1954).

See, T. J. J., 1909. On the cause of the remarkable circularity of the orbits of the planets and satellites and on the origin of the planetary system. *Astr. Nach.* No. 4308.

See, T. J. J., 1910. *Researches on the Evolution of Stellar Systems.* Lynn, Mass., pp. 375–376.

Seidelmann, P. K., 1971. A dynamical search for a trans-Plutonian planet. *Astron. J.* **76** 740.

Seidelmann, P. K., Marsden, B. G., and Giclas, H. L., 1972. Note on Brady's hypothetical trans-Plutonian planet. *Publ. Astron. Soc. Pacific* **84** 858 ff.

Sevin, M. E., 1946. Une planète transplutonienne. *Bull. Soc. Astron. de France* **60** 188. (In French.)

Shirokov, N. A., 1984. Estimate of the parameters of a trans-Plutonian planet. *VNII yader. geofiz. geokhim.* Moskva, 13 pp. (In Russian.)

Strubell, W., 1952. Existenmöglichkeit eines Transplutonische Planeten. *Die Sterne* **3** 70. (In German.)

Tomanov, V. P., 1986. On the problem of searching for new planets in the Solar System. *Astron. Tsirk.* No. 1444, 6–8. (In Russian.)

Tombaugh, C. W., 1960. Reminiscences of the discovery of Pluto. *Sky and Tel.* **19** 3.

Tombaugh, C. W., and Moore, P., 1981. *Out of the Darkness: the Planet Pluto.* Guildford and Harrisburg.

van Flandern, T. C., Kaplan, G. H., Pulkkinen, K. F., Santoro, E. J., and Seidelmann, P. K., 1980. The renewal of the trans-Neptunian planet search. *Bull. Amer. Astron. Soc.* **12** 830.

van Flandern, T. C., Pulkkinen, E. J., Santoro, E. J., Seidelmann, P. K., and Harrington, R. S., 1981. Perturbations of a trans-Neptunian planet. *Bull. Amer. Astron. Soc.* **13** 568.

Whitmire, D. P., and Matese, J. J., 1985. Periodic comet showers and Planet X. *Nature* **313** 36.

CHAPTER 13

Cruikshank, D. P., and Chapman, C., (Editors), 1995. *Neptune*. Univ. of Arizona Press.

Gehrels, T., and Matthews, M. S. (Editors), 1984. *Saturn*. Univ. of Arizona Press.

Hunt, G. E., and Moore, P., 1994. *Atlas of Neptune*. Cambridge Univ. Press.

Moore P., 1995. *Mission to the Planets*. Cassell, London.

Index

WILEY-PRAXIS SERIES IN ASTRONOMY AND ASTROPHYSICS
Forthcoming titles

THE VICTORIAN AMATEUR ASTRONOMER: Independent Astronomical Research in Britain 1820–1920
Allan Chapman, Wadham College, University of Oxford, UK

TOWARDS THE EDGE OF THE UNIVERSE: A Review of Modern Cosmology
Stuart G. Clark, Lecturer in Astronomy, University of Hertfordshire

LARGE-SCALE STRUCTURES IN THE UNIVERSE
Anthony P. Fairall, Professor of Astronomy, University of Cape Town, South Africa

MARS AND THE DEVELOPMENT OF LIFE, Second edition
Anders Hansson, Ph.D.

ASTEROIDS: Their Nature and Utilization, Second edition
Charles T. Kowal, Computer Sciences Corp., Space Telescope Science Institute, Baltimore, Maryland, USA

ACTIVE GALACTIC NUCLEI
Ian Robson, Director, James Clerk Maxwell Telescope, Head Joint Astronomy Centre, Hawaii, USA

ASTRONOMICAL OBSERVATIONS OF ANCIENT EAST ASIA
Richard Stephenson, Department of Physics, University of Durham, UK; Zhentao Xu, Purple Mountain Observatory, Academia Sinica, Nanjing, China; Yaotiao Tiang, Department of Astronomy, Nanjing University, China

EXPLORATION OF TERRESTRIAL PLANETS FROM SPACECRAFT, Second edition
Yuri Surkov, Chief of the Planetary Exploration Laboratory, Russian Academy of Sciences, Moscow, Russia